NATURAL POLYMERS FOR PHARMACEUTICAL APPLICATIONS

Volume 3: Animal-Derived Polymers

Natural Polymers for Pharmaceutical Applications, 3-volume set:

Natural Polymers for Pharmaceutical Applications
Volume 1: Plant-Derived Polymers

Natural Polymers for Pharmaceutical Applications
Volume 2: Marine- and Microbiologically Derived Polymers

Natural Polymers for Pharmaceutical Applications
Volume 3: Animal-Derived Polymers

NATURAL POLYMERS FOR PHARMACEUTICAL APPLICATIONS

Volume 3: Animal-Derived Polymers

Edited by

Amit Kumar Nayak, PhD
Md Saquib Hasnain, PhD
Dilipkumar Pal, PhD

Apple Academic Press Inc. | Apple Academic Press Inc.
3333 Mistwell Crescent | 1265 Goldenrod Circle NE
Oakville, ON L6L 0A2 | Palm Bay, Florida 32905
Canada | USA

© 2020 by Apple Academic Press, Inc.

Exclusive worldwide distribution by CRC Press, a member of Taylor & Francis Group

No claim to original U.S. Government works

Natural Polymers for Pharmaceutical Applications, Volume 3: Animal-Derived Polymers
International Standard Book Number-13: 978-1-77188-847-9 (Hardcover)
International Standard Book Number-13: 978-0-42932-835-0 (eBook)

Natural Polymers for Pharmaceutical Applications, 3-volume set
International Standard Book Number-13: 978-1-77188-844-8 (Hardcover)
International Standard Book Number-13: 978-0-42932-812-1 (eBook)

All rights reserved. No part of this work may be reprinted or reproduced or utilized in any form or by any electric, mechanical or other means, now known or hereafter invented, including photocopying and recording, or in any information storage or retrieval system, without permission in writing from the publisher or its distributor, except in the case of brief excerpts or quotations for use in reviews or critical articles.

This book contains information obtained from authentic and highly regarded sources. Reprinted material is quoted with permission and sources are indicated. Copyright for individual articles remains with the authors as indicated. A wide variety of references are listed. Reasonable efforts have been made to publish reliable data and information, but the authors, editors, and the publisher cannot assume responsibility for the validity of all materials or the consequences of their use. The authors, editors, and the publisher have attempted to trace the copyright holders of all material reproduced in this publication and apologize to copyright holders if permission to publish in this form has not been obtained. If any copyright material has not been acknowledged, please write and let us know so we may rectify in any future reprint.

Trademark Notice: Registered trademark of products or corporate names are used only for explanation and identification without intent to infringe.

Library and Archives Canada Cataloguing in Publication

Title: Natural polymers for pharmaceutical applications / edited by Amit Kumar Nayak, Md Saquib Hasnain, Dilipkumar Pal.

Names: Nayak, Amit Kumar, 1979- editor. | Hasnain, Md Saquib, 1984- editor. | Pal, Dilipkumar, 1971- editor.

Description: Includes bibliographical references and indexes.

Identifiers: Canadiana (print) 20190146095 | Canadiana (ebook) 20190146117 | ISBN 9781771888448 (set ; hardcover) | ISBN 9781771888455 (v. 1 ; hardcover) | ISBN 9781771888462 (v. 2 ; hardcover) | ISBN 9781771888479 (v. 3 ; hardcover) | ISBN 9780429328121 (set ; ebook) | ISBN 9780429328251 (v. 1 ; ebook) | ISBN 9780429328299 (v. 2 ; ebook) | ISBN 9780429328350 (v. 3 ; ebook)

Subjects: LCSH: Polymers in medicine. | LCSH: Biopolymers. | LCSH: Pharmaceutical technology.

Classification: LCC R857.P6 N38 2020 | DDC 615.1/9—dc23

CIP data on file with US Library of Congress

Apple Academic Press also publishes its books in a variety of electronic formats. Some content that appears in print may not be available in electronic format. For information about Apple Academic Press products, visit our website at **www.appleacademicpress.com** and the CRC Press website at **www.crcpress.com**

About the Editors

Amit Kumar Nayak, PhD

Amit Kumar Nayak, PhD, is currently working as an Associate Professor at Seemanta Institute of Pharmaceutical Sciences, Odisha, India. He has earned his PhD in Pharmaceutical Sciences from IFTM University, Moradabad, U.P., India. He has over 10 years of research experience in the field of pharmaceutics, especially in the development and characterization of polymeric composites, hydrogels, novel, and nanostructured drug delivery systems. Till date, he has authored over 120 publications in various high impact peer-reviewed journals and 34 book chapters to his credit. Overall, he has earned highly impressive publishing and cited record in Google Scholar (H-Index: 32, i10-Index: 80). He has been the permanent reviewer of many international journals of high repute. He also has participated and presented his research work at several conferences in India and is a life member of Association of Pharmaceutical Teachers of India (APTI).

Md Saquib Hasnain, PhD

Dr. Md Saquib Hasnain has over 6 years of research experience in the field of drug delivery and pharmaceutical formulation analyses, especially systematic development and characterization of diverse nanostructured drug delivery systems, controlled release drug delivery systems, bioenhanced drug delivery systems, nanomaterials and nanocomposites employing Quality by Design approaches as well as development and characterization of polymeric composites, formulation characterization and many more. Till date he has authored over 30 publications in various high impact peer-reviewed journals, 35 book chapters, 3 books and 1 Indian patent application to his credit. He is also serving as the reviewer of several prestigious journals. Overall, he has earned highly impressive publishing and cited record in Google Scholar (H-Index: 14). He has also participated and presented his research work at over 10 conferences in India, and abroad. He is also the member of scientific societies, i.e., Royal Society of Chemistry, Great Britain, International Association of Environmental and Analytical Chemistry, Switzerland and Swiss Chemical Society, Switzerland.

Dilipkumar Pal, PhD

Dilipkumar Pal, PhD, MPharm, Chartered Chemist, Post Doct (Australia) is an Associate Professor in the Department of Pharmaceutical Sciences, Guru Ghasidash Vishwavidyalaya (A Central University), Bilaspur, C.G., India. He received his master and PhD degree from Jadavpur University, Kolkata and performed postdoctoral research as "Endeavor Post-Doctoral Research Fellow" in University of Sydney, Australia. His areas of research interest include "Isolation, structure Elucidation and pharmacological evaluation of indigenous plants" and "natural biopolymers." He has published 164 full research papers in peer-reviewed reputed national and international scientific journals, having good impact factor and contributed 113 abstracts in different national and international conferences. He has written 1 book and 26 book chapters published by reputed international publishers. His research publications have acquired a highly remarkable cited record in Scopus and Google Scholar (H-Index: 35; i-10-index 82, total citations 3664 till date). Dr. Pal, in his 20 years research-oriented teaching profession, received 13 prestigious national and international professional awards also. He has guided 7 PhD and 39 master students for their dissertation thesis. He is the reviewer and Editorial Board member of 27 and 29 scientific journals, respectively. Dr. Pal has been working as the Editor-in-Chief of one good research journal also. He is the member and life member of 15 professional organizations.

Contents

Contributors .... *ix*

Abbreviations .... *xi*

Preface .... *xv*

1. Hyaluronic Acid (Hyaluronan): Pharmaceutical Applications 1

Amit Kumar Nayak, Mohammed Tahir Ansari, Dilipkumar Pal, and
Md Saquib Hasnain

2. Pharmaceutical Applications of Albumin 33

Suvadra Das and Partha Roy

3. Pharmaceutical Applications of Collagen 61

K. Sangeetha, A. V. Jisha Kumari, E. Radha, and P. N. Sudha

4. Pharmaceutical Applications of Gelatin 93

Gautam Singhvi, Vishal Girdhar, Shalini Patil, and Sunil Kumar Dubey

5. Pharmaceutical Applications of Chondroitin 117

Dilipkumar Pal, Amit Kumar Nayak, Supriyo Saha, and Md Saquib Hasnain

6. Biodegradability and Biocompatibility of Natural Polymers 133

Abul K. Mallik, Md Shahruzzaman, Md Sazedul Islam, Papia Haque,
and Mohammed Mizanur Rahman

Index *169*

Contributors

Mohammed Tahir Ansari
Department of Pharmaceutical Technology, Faculty of Pharmacy and Health Sciences, University Kuala Lumpur Royal College of Medicine Perak, Malaysia

Suvadra Das
Basic Science and Humanities Department, University of Engineering & Management, Kolkata, India

Sunil Kumar Dubey
Department of Pharmacy, Birla Institute of Technology & Science (BITS), Pilani, Rajasthan – 333031, India

Vishal Girdhar
Department of Pharmacy, Birla Institute of Technology & Science (BITS), Pilani, Rajasthan – 333031, India

Papia Haque
Department of Applied Chemistry and Chemical Engineering, Faculty of Engineering and Technology, University of Dhaka, Dhaka 1000, Bangladesh

Md Saquib Hasnain
Department of Pharmacy, Shri Venkateshwara University, NH-24, Rajabpur, Gajraula, Amroha–244236, U.P., India

Md Sazedul Islam
Department of Applied Chemistry and Chemical Engineering, Faculty of Engineering and Technology, University of Dhaka, Dhaka 1000, Bangladesh

A. V. Jisha Kumari
Department of Chemistry, Tagore Engineering College, Chennai, Tamil Nadu, India

Abul K. Mallik
Department of Applied Chemistry and Chemical Engineering, Faculty of Engineering and Technology, University of Dhaka, Dhaka 1000, Bangladesh

Amit Kumar Nayak
Department of Pharmaceutics, Seemanta Institute of Pharmaceutical Sciences, Mayurbhanj–757086, Odisha, India

Dilipkumar Pal
Department of Pharmaceutical Sciences, Guru Ghasidas Vishwavidyalaya, Koni, Bilaspur–495009, C.G., India

Shalini Patil
Department of Pharmacy, Birla Institute of Technology & Science (BITS), Pilani, Rajasthan – 333031, India

E. Radha
Biomaterials Research Lab, Department of Chemistry, D.K.M. College for Women (Autonomous), Vellore, Tamil Nadu, India

Mohammed Mizanur Rahman
Department of Applied Chemistry and Chemical Engineering, Faculty of Engineering and Technology, University of Dhaka, Dhaka 1000, Bangladesh

Partha Roy
Department of Pharmaceutical Technology, Adamas University, Kolkata, India

Supriyo Saha
School of Pharmaceutical Sciences and Technology, SardarBhagwan Singh University, Dehradun – 248161, Uttrakhand, India

K. Sangeetha
Biomaterials Research Lab, Department of Chemistry, D.K.M. College for Women (Autonomous), Vellore, Tamil Nadu, India

Md Shahruzzaman
Department of Applied Chemistry and Chemical Engineering, Faculty of Engineering and Technology, University of Dhaka, Dhaka 1000, Bangladesh

Gautam Singhvi
Department of Pharmacy, Birla Institute of Technology & Science (BITS), Pilani, Rajasthan – 333031, India

P. N. Sudha
Biomaterials Research Lab, Department of Chemistry, D.K.M. College for Women (Autonomous), Vellore, Tamil Nadu, India

Abbreviations

ADA-GEL	alginate dialdehyde-gelatin
APTI	Association of Pharmaceutical Teachers of India
BM	basement membrane
BSA	bovine serum albumin
BSE	bovine spongiform encephalopathy
CCC	collagen-based cell carrier
CDVA	corrected distance visual acuity
CMC	carboxymethyl cellulose
CPs	cocoa procyanidins
CS	chondroitin sulfate
CXL	corneal collagen cross-linking
DAC	drug affinity complex
DRG	dorsal root ganglia
DSC	differential scanning calorimetry
EC	ethyl cellulose
ECM	extracellular matrix
EDC/NHS	1-ethyl-(3-3 dimethylaminopropyl) carbodiimide hydrochloride/N-hydroxysuccinimide
EGCG	epigallocatechin gallate
EGF	epidermal growth factor
EGFR	epidermal growth factor receptor
EGFR2	epidermal growth factor receptor 2
EP	European Pharmacopeia
FFTC	Food and Fertilizer Technology Centre
FMD	foot-and-mouth disease
FTIR	Fourier transform infrared
GLP-1	glucagon-like peptide-1 agonist
GO-PEG	grapheme oxide–polyethylene glycol
GPMT	guinea pig maximization test
HA	hyaluronic acid
HA	hydroxyapatite
HAS	hyaluronan syntheses
HC	heavy chain

HEC	hydroxyethyl cellulose
HF	heart failure
HGC-27	human gastric cancer cell line
HGF	hepatocyte growth factor
HPC	hydroxypropyl cellulose
HPMC	hydroxypropylmethylcellulose
HSA	human serum albumin
ICL	implantable Collamer lens
ICTP	carboxy-terminal telopeptide of type-I collagen
IL-1ra	interleukin-1 receptor antagonist
IM	intramuscular
IV	intravenous
JP	Japanese Pharmacopeia
MC	methylcellulose
MEM	mammalian cell culture media
MHLW	Ministry of Health, Labor, and Welfare
MMPs	matrix metalloproteinases
MSC	mesenchymal stem cells
NaCMC	sodium carboxymethyl cellulose
NC	non-collageneous
NSAIDs	nonsteroidal anti-inflammatory drugs
PC-DAC	preformed conjugate-drug affinity complex
PCL	poly (ε-caprolactone)
PDS	pigment dispersion syndrome
PIIINP	procollagen type III N-terminal propeptide
PLGA	poly(lactide-co-glycolide)
PRK	photorefractive keratectomy
PTFE	polytetrafluoroethylene
PVP	polyvinylpyrrolidone
RSM	response surface methodology
SCA	starch cellulose acetate
scFv	single-chain antibody
SEM	scanning electron microscopy
SEVA-C	starch ethylene vinyl alcohol
SMCs	smooth muscle cells
SPARC	secreted protein, acidic, and rich in cysteine
TACS	tumor-associated collagen signatures
TEM	transmission electron microscopy

Abbreviations xiii

TGA	thermogravimetry analysis
TG-CXL	topography-guided corneal collagen cross-linking
TIMP 9	tissue inhibitors of matrix metalloproteinases
TNF-α	tumor necrosis factor-α
UDVA	uncorrected distance visual acuity
USP	United States Pharmacopeia
XRD	x-ray diffraction

Preface

In recent years, numerous animal-derived polymers have emerged as an attractive category of naturally derived polymers because of their advantageous physicochemical, chemical, as well as biological properties. The important biological properties of these animal-derived natural polymers are biocompatibility and biodegradation. These polymers are generally composed of repeated units of amino acids. Moreover, these polymers can be modified physically and/or chemically to improve their biomaterial properties. During the past few decades, a number of these animal-derived natural polymers have been explored and exploited as pharmaceutical excipients in various pharmaceutical dosage forms like microparticles, nanoparticles, ophthalmic preparations, gels, implants, etc. The commonly used animal-derived polymers used as pharmaceutical excipients are hyaluronic acid (HA) (hyaluronan), albumin, collagen, gelatin, chondroitin, etc.

This current volume of the book, *Natural Polymers for Pharmaceutical Applications, Volume 3: Animal-Derived Polymers,* contains six important chapters, that present the latest research updates on the animal-derived polymers for various pharmaceutical applications. The topics of the chapters of the current volume include but not limited to: Hyaluronic Acid (Hyaluronan): Pharmaceutical Applications; Pharmaceutical Applications of Albumin; Pharmaceutical Applications of Collagen; Pharmaceutical Applications of Gelatin; Pharmaceutical Applications of Chondroitin; and Biodegradability and Biocompatibility of Natural Polymers. This book mainly discusses the aforementioned topics along with an emphasis on the recent advances in the fields by the experts across the world.

We would like to thank all the authors of the chapters for providing timely and excellent contributions, the publisher Apple Academic Press (USA), and Sandra Sickels for her invaluable help in the editing process. We gratefully acknowledge the permissions to reproduce copyright materials from several sources. Finally, we would like to thank our family members, all respected teachers, friends, colleagues, and dear students for their continuous encouragement, inspiration, and moral support during the preparation of this book. Together with our contributing authors and the publishers, we will be extremely pleased if our efforts fulfill the needs of academicians, researchers, students, polymer engineers, and pharmaceutical formulators.

—Amit Kumar Nayak, PhD
Md Saquib Hasnain, PhD
Dilipkumar Pal, PhD

CHAPTER 1

Hyaluronic Acid (Hyaluronan): Pharmaceutical Applications

AMIT KUMAR NAYAK,[1] MOHAMMED TAHIR ANSARI,[2]
DILIPKUMAR PAL,[3] and MD SAQUIB HASNAIN[4]

[1]*Department of Pharmaceutics, Seemanta Institute of Pharmaceutical Sciences, Mayurbhanj–757086, Odisha, India*

[2]*Department of Pharmaceutical Technology, Faculty of Pharmacy and Health Sciences, University Kuala Lumpur Royal College of Medicine Perak, Malaysia*

[3]*Department of Pharmaceutical Sciences, Guru Ghasidas Vishwavidyalaya, Koni, Bilaspur–495009, C.G., India*

[4]*Department of Pharmacy, Shri Venkateshwara University, NH-24, Rajabpur, Gajraula, Amroha–244236, U.P., India*

ABSTRACT

Hyaluronic acid (HA) is a naturally occurring poly-anionic biomacro-molecular polysaccharide, discovered in bovine vitreous humor in 1934. Commercially HA is isolated either from animal sources, within the synovial fluid, umbilical cord, skin, and rooster comb, or from bacteria through a process of fermentation or direct isolation. Chemically, HA is composed of uronic acid and aminosugar. The disaccharides, D-glucuronic acid and d-N-acetylglucosamine, are linked together through alternating β-(1→4) and β-(1→3) glycosidic bonds. The average molecular weight of HA in human synovial fluid is 3–4 million Da, and HA purified from the human umbilical cord is 3,140,000 Da. The enzymatic degradation of the natural polymer- HA is catalyzed by hyaluronidase (hyase), β-d-glucuronidase, and β-N-acetylhexosaminidase. Many favorable properties like improved viscoelastic characteristics, biocompatibility, non-immunogenicity, and biodegradability lead HA as a useful polymeric excipient in various pharmaceutical applications, including drug delivery. The application has been extensively studied in parenteral, ophthalmic, nasal, and implantation drug delivery system.

1.1 INTRODUCTION

The utilization of the biopolymers in drug delivery system has been reported by various researchers (Behera et al., 2010; Bera et al., 2015a, b, c; Das et al., 2014; Hasnain et al., 2019; Maji et al., 2012; Malakar et al., 2012, 2014a, b, c; Nayak and Manna, 2011; Nayak and Das, 2018). The usage has been demonstrated for solid, liquid, and semi-solid dosage formulation and also used categorically for the design of modified release drug delivery systems. It is available as a natural polymer but can also be synthesized in the laboratory (Adhikari et al., 2010; Das et al., 2013, 2017; Jana et al., 2015a, b; Malakar et al., 2013; Malakar and Nayak, 2013; Nayak and Malakar, 2010; Nayak et al., 2011, 2013a, 2018a, b, c, d; Verma et al., 2017). The close proximity of natural based polymer with the extracellular materials minimize the possible hypersensitivity reactions, inflammatory responses and toxicity effects which are very prominent with synthetic polymers (Hasnain et al., 2010; Nayak and Pal, 2012; Nayak et al., 2018c). The auto-degradation of most natural polymers due to the normal enzymatic activity of body presents it to be excellent candidates for a variety of application in medical and biomedical avenues, such as a polymer or a carrier in drug delivery system (Nayak et al., 2018d). Moreover, the application of natural polymers in medical and biomedical field has attracted attention as it is easily available, non-toxic, non-irritant, easily degradable, ease in chemical modification, abundance, and most of them are compatible with human body (Hasnain and Nayak, 2018a; Hasnain et al., 2018a; Nayak et al., 2010a, b; Pal and Nayak, 2015a; Pal et al., 2010).

Previous studies have shown extensive usage of natural biopolymers as an effective pharmaceutical excipient for the design of various drug delivery system (Nayak and Pal, 2016a,b; Pal and Nayak, 2017). The use of gum Arabica (Nayak et al., 2012a), sodium alginate (Sant et al., 2011; Hasnain et al., 2016, 2018a; Hasnain and Nayak, 2018b; Jana et al., 2014a,

Hyaluronic Acid (Hyaluronan): Pharmaceutical Applications

2015a; Nayak et al., 2018e; Pal and Nayak, 2015b), chitosan (Nayak and Pal, 2015a; Pal et al., 2018; Ray et al., 2018), gellan gum (Jana et al., 2013; Nayak et al., 2014a; Nayak and Pal, 2014), pectin (Nayak et al., 2013b, c, 2014b), guar gum (Ali et al., 2010; Soumya et al., 2010), agar gum (Saleem et al., 2008), locust bean gum (Dionisio and Grenha, 2012), sterculia gum (Guru et al., 2013; Nayak and Pal, 2015b), okra gum (Nayak et al., 2018a; Sinha et al., 2015a, b), dillenia gum (Kuotsu and Bandyopadhyay, 2007), tamarind seed polysaccharide (Nayak et al., 2012b, 2014b, c, 2016; Nayak and Pal, 2013a, 2017a, b; Nayak, 2016), fenugreek seed polysaccharide (Nayak et al., 2012c, 2013d), linseed polysaccharide (Hasnain et al., 2018b), cashew gum (Hasnain et al., 2017a, b, 2018a), gum odina (Jena et al., 2018), isabghula polysaccharide (Guru et al., 2017; Tahir et al., 2010; Nayak et al., 2010a, 2013e), starches (Malakar and Nayak, 2013; Malakar et al., 2013; Nayak and Pal, 2013b, 2017c; Nayak et al., 2013e, 2014d), hyaluronic acid (HA) (Brown and Jones, 2005; Huang and Huang, 2018), gelatin (Foox and Zilberman, 2015), albumin (Jana et al., 2013, 2014b), etc., have been reported in literature. Amongst these natural biopolymers, HA is a naturally occurring poly-anionic biomacromolecular polysaccharide, discovered in bovine vitreous humor in 1934. It has an important biological function in bacteria and higher animals, including humans. HA is an abundant polysaccharide, exists *in vivo* as a polyanion and also is a major constituent of the synovial fluid, it is distributed widely in vertebrate connective tissues especially skin where it has a protective, structure stabilizing and shock-absorbing role. It was also was described as a "goo" molecule due to its abundance in extracellular material (Brown and Jones, 2005; Hargittai et al., 2011; Huang and Huang, 2018; Meyer and Palmer, 1936; Toole, 2000).

The primary biomedical use of HA was as vitreous substitution/ replacement during an eye surgery. HA was first isolated from the umbilical cord and, later also from rooster combs (Hargittai et al., 2011). The structural/biological characteristics of this polysaccharide were explored more deeply in several laboratories. It was surmised that HA is viscoelastic in nature, extensive studies on the chemical and physicochemical properties of HA envisaged its supreme biocompatibility, non-irritancy, and non-immunogenicity. This has allowed its usage in various clinical applications, such as: a surgical aid in eye surgery, regeneration of surgical wounds and eventually assists in healing, a supplement for synovial fluid in arthritis. Furthermore, it has been proved to be ideal biopolymers for cosmetic, and pharmaceutical applications (Baier Leach and Schmidt, 2004; Bot et al., 2008; Brown and Jones, 2005; Huang and Huang, 2018; Necas et al., 2008). Studies have reported successful application of HA as a drug delivery agent, carrier or polymer for various routes of administration such as ophthalmic, nasal, pulmonary, parenteral, and topical (Huang and Huang, 2018; Ossipov, 2010; Widjaja et al., 2013). This chapter relates all the useful and a complete discourse on the pharmaceutical applications of HA. Besides this, some significant features of HA like source, extraction, chemical composition, and characteristics are also discussed in brief.

1.2 SYNTHESIS

Hyaluronate, a salt form of HA is found in human. It is predominantly present in soft connective tissues, such as umbilical cord, skin synovial fluid, and vitreous humor. Substantial concentrations of HA as hyaluronate are also present in the brain, lung, muscle tissues, and kidney (Brown and Jones, 2005). It is understood that the cellular synthesis of HA is a unique and highly controlled process, which is primarily located in Golgi bodies (Necas et al., 2008). The synthesis of HA is catalyzed by an enzyme hyaluronan synthases (HAS). HAS are classified as integral membrane proteins. Integral membrane proteins are found mostly embedded or associated in the cell membrane or plasma membrane. HAS have been classified as HAS1, HAS2, and HAS3 in vertebrates (Garg and Hales, 2004; Lee and Spicer, 2000). The function of HAS enzymes is to synthesize linear and long polymers of hyaluronan, a disaccharide structure with an addition of glucuronic acid and N-acetyl glucosamine (Necas et al., 2008).

Commercially HA is isolated either from animal sources, within the synovial fluid, umbilical cord, skin, and rooster comb, or from bacteria through a process of fermentation or direct isolation (Necas et al., 2008). A process has been developed to synthesize HA genetically from a bacteria *Bacillus subtilis*, and this genetically modified process produces HA that is fit for the human-use product (Menetrey et al., 2012).

1.3 CHEMICAL COMPOSITION AND CHARACTERISTICS

Chemically, HA is composed of uronic acid and aminosugar. The disaccharides, D-glucuronic acid and d-N-acetyl glucosamine, are linked together through alternating β-(1→4) and β-(1→3) glycosidic bonds (Figure 1.1) (Necas et al., 2008). The structure of the disaccharide is energetically very stable as both sugars are spatially related to glucose. The stearic arrangement of the functional group make the structure more stable as the carboxylate moiety, the hydroxyl group, and the anomeric carbon which are the bulky groups are placed at sterically placed at equatorial positions while the hydrogen atoms or the lighter groups are at sterically favorable axial positions (Necas et al., 2008). The number disaccharides in a complete HA molecule can reach 10,000 or up to 25,000 units, with a molecular mass of ~4 million Da (Cowman and Matsuoka, 2005). The average molecular weight of HA in human synovial fluid is 3–4 million Da, and HA purified from the human umbilical cord is 3,140,000 Da (Garg and Hales, 2004; Necas et al., 2008).

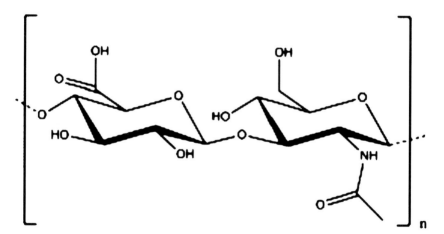

FIGURE 1.1 Chemical structure of HA.

The interaction of physiological solution stiffens the backbone of an HA molecule due to the combination of the disaccharide unit and, the presence of internal hydrogen bonding, enabling a coiled structure that traps approximately 1000 times its weight in water (Necas et al., 2008). The non-polar hydrophobic face is characterized by the presence of axial hydrogen atoms, and a relatively hydrophilic face is equipped with the polar-equatorial side chains to form, thereby creating a twisting ribbon structure. HA solutions show varied rheological properties and are very hydrophilic and extremely lubricious (Cowman and Matsuoka, 2005).

The unusual rheological behavior is exhibited due to the formation of an expanded random coil. This is due to the entanglement of chains at a very low concentration. At higher concentrations, solutions have an extremely high but shear-dependent viscosity. 1% HA solution behaves like jelly and under stress may easily move through a needle. Owing to this behavior, it has, therefore classified as "pseudo-plastic" substance (Necas et al., 2008). This rheological behavior makes it viable to be used as a lubricant. Hence, its usage as a replacement fluid in joints and as vitreous humor have been accepted. The rheological behavior has extended its application in reducing the postoperative adhesion formation following abdominal and orthopedic surgery (Necas et al., 2008).

The enzymatic degradation of the natural polymer- HA is catalyzed by hyaluronidase (hyase), β-d-glucuronidase, and β-N-acetylhexosaminidase. These degrading enzymes are found intracellularly and in the serum of all mammals (Baier Leach and Schmidt, 2004). Hyase cleaves high molecular weight HA into smaller oligosaccharides, and the terminal non-reducing sugars are fragmented by β-d-glucuronidase and β-N-acetylhexosaminidase (Necas et al., 2008).

Pro-angiogenic properties have been reported by the degraded products of HA (Baier Leach and Schmidt, 2004; Huang and Huang, 2018). Tissue permeability is enhanced by lowering the viscosity of HA, which is achieved by catalytic hydrolysis of HA by hyaluronidase. Hence, it has found its application in drug delivery system as it is capable of increasing the rate of drug dispersion in the body and also, assists in the delivery to the target tissue. The most common use of such an application is during eye surgery (Girish and Kemparaju, 2006; Mio and Stern, 2002).

The hydrophilic nature of HA enables it to form a stable with hydrazide derivatives (Drury and Mooney, 2003). The hydrogels transformed HA has profound physiological such as cells, matrix, and tissue water regulation, filling properties during surgery, lubrication for many drug delivery system and joints, many number of macromolecular functions (Hasnain et al., 2010; Kurisawa et al., 2005). The lubricant activity is enriched with shock absorbing capacity in the synovial fluid due to its hydrogel formation (Nishinari and Takahashi, 2003). It can also be used for its antioxidant and scavenging activity in wound sites, and hence also modulating inflammation (Hasnain et al., 2010). These properties also allow HA to be used for wound dressing applications (Nair and Laurencin, 2006). The bacteriostatic property has also allowed it to be mandated for wound dressing application. Recently the application of HA has been investigated for gene delivery (Huang and Huang, 2018; Kurisawa et al., 2005).

1.4 PHARMACEUTICAL APPLICATIONS

Various favorable properties like improved viscoelastic characteristics, biocompatibility, non-immunogenicity, and biodegradability lead HA as a useful polymeric excipient in various pharmaceutical applications, including drug delivery. The application of has been extensively studied in parenteral (Sakurai et al., 1997), ophthalmic (Camber and Edman, 1989; Camber et al., 1987), nasal (Lim et al., 2002; Morimoto et al., 1991) and implantation drug delivery system (Surini, 2003). The application is mostly due to its viscoelastic behavior, which makes it a good candidate to possess the mucoadhesive property. This will allow the drug delivery system to modify the *in vivo* release/absorption rate of the therapeutic agent (Takayama et al., 1990).

1.4.1 TOPICAL DRUG DELIVERY

HA in a topical formulation offers clear and unique potential in the delivery and localization of drugs to the skin (Brown and Jones, 2005; Brown et al., 1995). HA-based gel of diclofenac sodium branded as 'Solarize' for the topical treatment of actinic keratosis, a skin infection common in western countries as was approved for human use after a successful randomized clinical trial in Canada, USA, and European countries (Jarvis and Figgitt, 2003). The gel is well tolerated, safe, and efficacious and provides an attractive, cost-effective alternative to cryoablation, curettage or dermabrasion, or treatment with 5-fluorouracil (Brown and Jones, 2005). Research studies computed that 3% diclofenac in 2.5% HA gel vehicle is well tolerated without any major hypersensitive or immunological reaction in randomized, double-blind, controlled clinical studies (McEwan and Smith, 1997; Rivers, 1997; Wolf et al., 2001). Majority of the studies or clinical trial concluded that most of the individuals tolerated the formulated gel with few of the individuals reporting minor adverse effects like dry skin, skin rash, and exfoliation locally on the application site. The HA gel was proved to be safe and non-fatal by all the clinical studies. Withdrawal symptoms were occasionally reported during the trials. Hence, it was suggested and concluded that 3% diclofenac in 2.5% HA gel appears to be most suitable for any lesions on face, forehead, or skin. It was also suggested that the efficacy of the formulation increases with the duration of the formulation of individuals. Lesion was cured completely with the gel when it was applied twice for 2 to 3 months. HA-based diclofenac gel proved to an effective replacement to costly treatment module which such as 5-fluorouracil gel, which also was not advised due to major adverse effect.

Brown et al., (2001) have further confirmed the major benefits of HA-based topical drug delivery system. The study concluded that HA significantly enhances the partitioning of drug diclofenac into the skin and also increases its epidermal localization and retention. The HA gel was compared with aqueous control, chondroitin sulfate, and sodium carboxymethyl cellulose (NaCMC) for localization and retention control. HA-based gel figured higher retention than any gel-based system. Many other studies also supported the retention behavior of HA-based gel system, which allowed to control the release of the drug by forming a depot or reservoir of the drug in the epidermis (McEwan and Smith, 1997; Wolf et al., 2001). Moreover, a study by Lin and Maibach (1996) depicted that demonstrated HA-based gel delivered twice as much diclofenac to the epidermis over 24 h when compared with an aqueous control and sodium CMC. HA-based gel delivery system showed similar effects with ibuprofen (Brown et al., 2001) clindamycin phosphate (Sk, 2000) and cyclosporine (Brown, 1995). Some other examples of HA-based topical transdermal drug delivery systems are presented in Table 1.1.

1.4.2 OPHTHALMIC DRUG DELIVERY

HA has been known to extend the pre-corneal residence time, and hence, it is an important constituent of eye drops. It also has the capacity to increase the viscosity and the mucoadhesive properties of the eye drops (Camber and Edman, 1989; Camber et al., 1987; Salzillo et al., 2016). A study aiming to maximize the viscosity and mucoadhesiveness of HA-based preparations were performed by Salzillo et al., (2016). He demonstrated the effects of polymer chain length and concentration of HA on the effectiveness of eye drops. It was deduced through the study that 0.3 (wt.%) for an 1100 kDa HA up to 1.0 (wt.%) for a 250 kDa HA can be used in eye formulations. The rheological properties and mucoadhesion of HA-based

TABLE 1.1 Some Examples of HA-Based Topical Transdermal Drug Delivery Systems

HA-Based Topical Transdermal Drug Delivery Systems	Drugs	References
Bioadhesive transdermal device from chitosan and HA	Lidocaine	Anirudhan et al., 2016
HA-based liposomes used as bioadhesive carriers for topical drug-delivery in wound-healing	Epidermal growth factor (EGF)	Yerushalmi et al., 1994
Chitosan/hyaluronan film for transdermal delivery	Thiocolchicoside	Bigucci et al., 2015
Nanocrystals hydrogels formulation of HA for topical application of drug	Baicalin	Wei et al., 2018
HA-containing ethosomes as a potential carrier for transdermal drug delivery	Rhodamine B	Xie et al., 2018
Nanoparticle-based hyaluronate gel containing retinyl palmitate for wound healing	Retinyl palmitate	Esposito et al., 2018
HA-based nano-emulsion for inhibition of keloid fibroblast	10, 11-methylenedioxy camptothecin	Gao et al., 2014
HA-hydroxyethyl cellulose hydrogels for transdermal delivery	Isoliquiritigenin	Kong et al., 2016
HA modified nanostructured lipid carriers for transdermal delivery	Bupivacaine	Yue et al., 2018
pH- and temperature-sensitive double cross-linked interpenetrating polymer network hydrogels based on HA/poly (N-isopropyl acrylamide) for transdermal delivery	Luteolin	Kim et al., 2018
Novel microneedle arrays fabricated from HA for transdermal delivery	Insulin	Liu et al., 2012

eye formulation were superior to the commercial products in the market. HA-based eye formulation was also capable of entrusting protection corneal porcine epithelial cells from dehydration. As the concentration of the HA polymer was increased in the eye formulation, the bioavailability of the drug was also increased, suggesting that HA was much better than the available commercial formulations. This finding is valuable in the tuning of formulations, including HA, extending the precorneal residence time of the active ingredient (Salzillo et al., 2016).

The effect of a salt form of HA, sodium hyaluronate on the miosis induced by pilocarpine in rabbits was studied by Camber. Increased duration of miosis was evident from the extended AUC curve of the pilocarpine due to the addition of 0.2 and 0.75% sodium hyaluronate to 1% pilocarpine hydrochloride. The study proved that pilocarpine formulation composed of sodium hyaluronate enhanced the retention parameters without showing any adverse reaction and thereby increasing the bioavailability of the drug (Camber et al., 1987). Similar results were displayed by sodium hyaluronate on corneal residence time in rabbits. Addition of 0.125% of sodium hyaluronate to radiolabeled 3H-pilocarpine HCl solution resulted in increased retention of a radioisotope of hydrogen in tear fluid and a 2-fold increase in pilocarpine concentration in the cornea. The study on miosis effects also produced similar results, an increase in the concentration of sodium hyaluronate increased the corneal retention time and thereby increasing the miotic effect of the drug. This might indicate that other physicochemical properties of sodium hyaluronate influence drug bioavailability (Camber and Edman, 1989). In a research, Aragona (2002) studied sodium hyaluronate-based ophthalmic drops having various osmolarity. They found that these ophthalmic were useful for the use in dry eye treatment in case of Sjögren's syndrome. The use of a formulation with pronounced hypotonicity showed better effects on corneo conjunctival epithelium than the isotonic solution. Some other examples of HA-based ophthalmic drug delivery systems are presented in Table 1.2.

1.4.3 NASAL AND PULMONARY DRUG DELIVERY

HA has also been exploited for both nasal and pulmonary drug delivery applications. The nasal and pulmonary drug delivery approaches were found effective to deliver a wide range of drugs to treat nasal as well as pulmonary diseases (Vishvkarma et al., 2010). In a research, Morimoto

TABLE 1.2 Some Examples of HA-Based Ophthalmic Drug Delivery Systems

HA-Based Ophthalmic Drug Delivery Systems	Drugs	References
0.5% pilocarpine with sodium hyaluronate	Pilocarpine	Xu et al., 2004
HA-based ophthalmic drug delivery	Methylprednisolone	Kyyrönen et al., 1992
HA-based vehicle for treatment of dry eye	Cortisol phosphate	Rolando and Vagge, 2017
High molecular weight and low molecular weight fractions of sodium hyaluronate-based ophthalmic drug delivery systems	Pilocarpine	Saettone et al., 1991
Sodium hyaluronate-based for intraocular drug delivery systems	Gentamicin	Moreira et al., 1991a, b
Cross-linked HA films for ocular therapy	Dexamethasone	Calles et al., 2016
Hyaluronan-timolol ionic complexes extended release ophthalmic formulations	Timolol	Battistini et al., 2017
Thermo-sensitive hydrogel-based on poly (N-isopropylacrylamide)/HA for ophthalmic drug delivery	Ketoconazole	Zhu et al., 2018
HA and β-cyclodextrins films	Dexamethasone	Fiorica et al., 2017
HA-coated niosomes facilitate ocular drug delivery	Tacrolimus	Zeng et al., 2016
Hyalugel-integrated liposomes as a novel ocular nanosized drug delivery system	Fluconazole	Moustafa et al., 2017
HA modified MPEG-b-PAE block copolymer aqueous micelles	Genistein	Li et al., 2018
HA-modified lipid-polymer hybrid nanoparticles as an efficient ocular delivery platform	Moxifloxacin HCl	Liu et al., 2018
Cross-linked ocular inserts of sodium hyaluronan and hydroxypropyl-β-cyclodextrin	Cyclosporine	Grimaudo et al., 2018
Micelles of HA derivatives for the potential treatment of neovascular ocular diseases	Imatinib	Bongiovi et al., 2018

et al., (1991) investigated the influences of sodium hyaluronate solutions for nasal absorption of drugs. In this work, the absorptions of vasopressin through nasal route and its analog, 1-deamino-8-D-arginine vasopressin from the viscous solutions of sodium hyaluronate were measured and analyzed. Sodium hyaluronate solutions having molecular weight more than 3×10^5 Da improved absorption of vasopressin through the nasal route, whereas sodium hyaluronate solutions of molecular weight 5.5×10^4 Da were considered not useful. It was evident that the improved absorption of vasopressin and 1-deamino-8-D-arginine vasopressin through the nasal route was governed by the concentration of HA in the range of 0–1.5% (w/v). It was also surmised that the absorption of vasopressin via the nasal route was more at lower pH. Sodium hyaluronate solutions (molecular weight 1.4×10^6 Da and 2×10^6 Da) was able to enhance the bioavailability of vasopressin and 1-deamino-8-D-arginine vasopressin by 2and 1.6, respectively, through the nasal route, when compared to the administration of vasopressin and 1-deamino-8-D-arginine vasopressin through the nasal route using a buffer solutions (pH 7.0). The effects of sodium hyaluronate solution of molecular weight 1.4×10^6 Da did not disturb the ciliary beat frequency of rabbit nasal mucosal membranes, *in vitro*. Hence, it was concluded that the sodium hyaluronate solution can be adjudged as a useful vehicle for nasal delivery of vasopressin and 1-deamino-8-D-arginine vasopressin (Morimoto et al., 1991). Lim et al., prepared a novel HA/chitosan microparticulate system by solvent evaporation technique for nasal drug delivery. They loaded gentamicin (an aminoglycoside antibiotic) within these HA/chitosan microparticulate and the efficacy of gentamicin-loaded microparticulates was evaluated in rabbits, *in vivo*. These gentamicin-loaded microparticulates were administered through the nasal route by the use of an insufflator. A similar study was conducted with gentamicin in rabbits, where gentamicin was administered as powder, solution, intravenous (IV) injections, and intramuscular (IM) injections. Fluorescence polarization immunoassay was used to quantify the level of gentamicin in serum. The nasal solution bioavailability was measured as 1.1%, and the dry powder bioavailability was equated only 2.1% of the absolute bioavailability. However, a different microparticulate system composed of chitosan and HA/chitosan was able to increase the bioavailability of gentamicin to 31.4 and 42.9% respectively when compared to the absolute bioavailability. Previous literature has confirmed the mucoadhesive properties of chitosan/HA microparticulate system. The chitosan/HA system was also able to prolong the drug release of the drug. These findings suggest that the gentamicin bioavailability can be improved by using the HA-based chitosan microparticulate system (Lim et al., 2002). In another research, Di Cicco evaluated the efficiency and tolerability of a nasal spray formulation made of HA. Within the nasal spray, tobramycin was loaded, and tobramycin containing nasal spray formulation was tested in the cystic fibrosis patients with bacterial rhinosinusitis, and the result indicated the efficiency of the HA-based nasal delivery (Di Cicco et al., 2014).

Morimoto et al., (2001) evaluated pulmonary delivery of rh-insulin through low viscous sodium hyaluronate system. They also assessed the influences of various concentrations and pH of low viscous sodium hyaluronate system on the pulmonary absorption in rats. The sodium hyaluronate of a molecular weight of 2140 kDa solutions at 0.1% and 0.2% w/v at pH 7.0 significantly improved the therapeutic availability of rh-insulin, when compared to the rh-insulin aqueous solution at pH 7.0. The absorption of rh-insulin was greater for 0.1 w/v concentration of sodium hyaluronate was than that of 0.2% w/v concentration of sodium hyaluronate. The absorbing capacity of rh-insulin was insensitive the varied molecular weight of sodium hyaluronate (Morimoto et al., 2001). In another research, Surendrakumar et al., (2003) investigated the sodium hyaluronate-based dry powder containing insulin for pulmonary delivery. In this work, the systemic levels of insulin and the corresponding levels of glucose were measured after the administration of the sodium hyaluronate-based microparticles to the lungs in male Beagle dogs. Excess zinc ions or hydroxypropyl cellulose (HPC) were added to modify the release kinetics. It was found that formulations of insulin containing HA were able to prolong the terminal half-life and mean residence time when compared to spray-dried pure insulin. Zinc ions also prolonged the mean residence time by 9 folds, AUC/dose was increased by 2.5 folds, and T_{max} was superseded by a factor of 3, as compared to pure insulin made powder by a spray dried process. Similarly, addition of HPC also amended the mean residence time by 7 folds, the AUC/dose by 5 times and T_{max} was bettered by 3 times when compared as compared to pure insulin made powder by a spray dried process. The results establish the importance of HA-based dry powder solid formulation, which can be dispensed controlled release insulin via a pulmonary or a nasal route (Surendrakumar et al., 2003).

1.4.4 IMPLANTABLE DRUG DELIVERY

HA has also been employed as an implantable biomaterial in various implantable drug deliveries. An implant controlled-release system for protein drug delivery based on a polyion complex device composed The requirements to the greatest amount of the polyion complex formed was envisaged through a study it was concluded that the polyion complex was formed at chitosan: HA ratio of 3 to 7 at pH 3.5. This ratio was used to formulate the chitosan-HA pellets, and the drug release pattern of insulin from the chitosan-HA pellets was studied. The study concluded that the drug release of insulin was modified from the pellets, and it was further maneuvered by the change in the ratio of chitosan and HA polymer (Surini, 2003). In another work, Bettin et al., (2016) evaluated long-term efficacy and safety of deep sclerectomy augmented with mitomycin C and injectable cross-linked HA-based implant in medically refractory glaucoma patients. The study included 96 eyes of 83 consecutive patients with open-angle glaucoma undergoing mitomycin C with injectable cross-linked HA implant. Mean follow-up was 28.6 ± 20.0 months. Variables analyzed were: intraocular pressure, acuity, the best-corrected visual, mean number of antiglaucomatous drugs, execution of postoperative maneuvers such as goniopuncture, and bleb needling. Tonometric success was defined by two different thresholds, specifically intraocular pressure ≤ 21 mm Hg (criterion A) and ≤15 mm Hg (criterion B). The procedure was defined as a qualified success if reached with medication and as a complete success if reached without (Bettin et al., 2016).

1.4.5 DELIVERY OF ANTICANCER DRUGS

Studies have proved that HA has the capacity of binding to the receptors present on the surface of cancer cells. Owing to its compatibility and biodegradability, HA has made immense progress in targeting the cancer cells in the human body (Huang and Huang, 2018; Rosso et al., 2013). Research studies have shown its usage as a drug carrier and conjugate, which is capable of producing controlled release and also envisages drug targeting to various pathological areas in the human body. This unique application allows the researchers to realize the importance of the timed and directional release of active ingredients. But still, the major issue is that HA is readily degradable in the human body (Bot et al., 2008; Chen et al., 2014;

Huang and Huang, 2018). Hence, it is essential to that nitroxide-containing substance or a hyaluronidase inhibitor to be added in order to prevent the degradation of HA (Huang and Huang, 2018; Sung et al., 2014).

A double-targeted ternary complex containing HA and linked polyethyleneimine-dexamethasone/DNA having a nucleo-shell structure with HA and DNA was synthesized to evaluate the efficacy against B16–F10 tumor cells. The HA/polyethyleneimine-dexamethasone/DNA ternary complex exhibited low toxicity and high transfection efficiency in the targeted tumor cells. The ternary complex promotes cellular uptake and DNA nuclear translocation. Studies in tumor-bearing mice also revealed that the ternary complex exhibited an obvious anti-inflammatory activity and profound tumor growth inhibition (Fan et al., 2012). The drug conjugates act as prodrugs, which are synthesized or prepared by bonding antitumor drugs to HA. The bonds formed are very strong and not easily broken, but they are fragmented only through hydrolysis or enzymolysis at the target site and thus assist in targeting of the tumor cells. It has also been diagnosed that the drug conjugates containing HA have improved drug solubility, and subsequently affected the drug distribution and half-life. The complex conjugate favors the accumulation of the drug in the tumor cells by enhancing the osmotic retention effect (Fan et al., 2015). Fan et al., (2015) also formulated a pH-responsive HA-based complex system of cisplatin, which exhibited enhanced antitumor effect against targeted tumor cells. HA-paclitaxel conjugate has also been reported by synthesized to reduce the toxicity of taxanes and improve the antitumor activity by Galer et al., (2011). The drug conjugate exhibited a growth-inhibiting effect on head and neck squamous cell carcinoma cell lines OSC-19 and HN5. Increased survival rate was due to effective inhibition of tumor cells by HA-paclitaxel drug conjugate was exhibited in mice, certifying the enhanced antitumor activity due to the presence of HA. It is envisaged that different cross-linking agents could regulate the drug release rate and may provide more drug HA conjugates with varied properties. Xin et al., (2009) used amino acid as a crosslinking agent to make a conjugate of paclitaxel and HA. The conjugate was linked using the carboxyl group of amino acid and the hydroxyl group of paclitaxel. This bi conjugate was linked to the carboxyl group of HA via the carboxyl group of amino acid to form a ternary drug conjugate. They used amino acid as a cross-linker to link paclitaxel to HA. The presence of amino acids ensured that paclitaxel was released faster from the conjugate. Cell line studies showed that ternary complexes also enhanced the toxicity of breast cancer cells, leaving the cell cycle in G(2)/M stage (Xin et al., 2009). Some other examples of HA-based anticancer drug delivery systems are presented in Table 1.3.

1.4.6 DELIVERY OF PROTEIN AND PEPTIDE DRUGS

Recent years, numerous biopolymers are being utilized for the effective delivery of various kinds of protein and peptide drugs (Nayak, 2010). HA is also found capable of delivering the protein and peptide drugs effectively; and, these can be used to treat and management of various diseases. The application of HA as a new protein and peptide drug carrier has opened a new avenue for research. The delivery mechanism for proteins and peptides using HA is different than PEGylation. Individual HA chain may conjugate with different peptide chain molecules and thus enabling a polypeptide drugs structure, which can exert multiple effects (Jiang et al., 2012). HA is a suitable polymer, which can be used to prolong the release of protein drugs. The mechanism of release can be visualized in an aqueous medium. Hydrogels formulation using HA has been reported in the literature for encapsulating and prolonging the release of protein drugs (Hirakura et al., 2010). Highly crosslinked HA microgel network structure is required to sustain or prolong the release of protein drugs. Protein drugs of particle size 3 to 15 nm can be formulated in a hydrogel mesh size of 5 and 25 nm, and the mesh would be able to encapsulate and sustain the release of protein drug through diffusion process. Sustained release erythropoietin was enabled at different PKa in a crosslinked HA-based hydrogel, and crosslinking was cultured using the hydrazide group of HA-ADH and the amino group of protein drug (Motokawa et al., 2006). Hahn et al., (2007) reported an *in vitro* and *in vivo* controlled released EPO using an EPO-ThioHA (Thiohyaluronic acid) particulate hydrogel. He formulated an HA-based microhydrogel, EPO loading was catalyzed by sodium tetrathionate. The addition of sodium tetrathionate modulated the gelation time of HA micro-hydrogel from 1 day to 30 minutes. The *in vitro* and *in vivo* release was modulated due to the formation of disulfide bond due to the presence of thioHA (Hahn et al., 2007). Some other examples of HA-based protein and peptide drug delivery systems are presented in Table 1.4.

TABLE 1.3 Some Examples of HA-Based Anticancer Drug Delivery Systems

HA-Based Anticancer Drug Delivery Systems	Drugs	References
Anticancer drug-clusters coated with HA as selective tumor-targeted nanovectors	Paclitaxel	Rivkin et al., 2010
HA nanoparticles based on ion complex formation	Cisplatin	Jeong et al., 2008
HA-based hydrosoluble bioconjugate for treatment of superficial bladder cancer	Paclitaxel	Rosato et al., 2006
Dual targeting folate-conjugated HA polymeric micelles	Paclitaxel	Liu et al., 2011
Self-assembled nanoparticles based on HA-ceramide and Pluronic® for tumor-targeted delivery	Docetaxel	Cho et al., 2011
TiO$_2$-HA modified nanoparticles for neoadjuvant chemotherapy of ovarian cancer	Cisplatin	Liu et al., 2015
Long-circulating HA targeted nano-liposomes increases its antitumor activity in three mice tumor models	Mitomycin C	Peer and Margalit, 2003
Paclitaxel-HA bio-conjugate for targeting ovarian cancer	Paclitaxel	Banzato et al., 2008
HA-based nanoprodrug with the cytosolic mode of activation for targeting cancer	Camptothecin	Yang et al., 2013
Hollow HA particles by competition between adhesive and cohesive properties	Catechol	Lee et al., 2017
Conjugating an anticancer drug onto thiolated HA by acid liable hydrazone linkage for its gelation and dual stimuli-response release	Doxorubicin HCl	Fu et al., 2015
HA-incorporated nanostructure of a peptide-drug amphiphile for targeted anticancer drug delivery	Camptothecin	Choi et al., 2016
HA-conjugated apoferritin nanocages for lung cancer targeted drug delivery.	Daunomycin	Luo et al., 2015
Ovarian cancer targeted HA-based nanoparticle system	Paclitaxel	Wang and Jia, 2015
HA-decorated polymeric nanoparticles	Tamoxifen and Docetaxel	Zhu et al., 2017
Multifunctional HA modified graphene oxide loaded with the anticancer drug for overcoming drug resistance in cancer	Mitoxantrone	Hou et al., 2015
Nanohybrid magnetic liposome functionalized with HA for enhanced cellular uptake and near-infrared-triggered drug release	Docetaxel	Nguyen et al., 2017
HA-decorated graphene oxide nanohybrids as nanocarriers for targeted and pH-responsive anticancer drug delivery	Doxorubicin	Song et al., 2014

TABLE 1.4 Some Examples of HA-Based Protein and Peptide Drug Delivery Systems

HA-Based Protein and Peptide Drug Delivery Systems	Drugs	References
Low viscosity sodium hyaluronate preparation on the pulmonary absorption of rh-insulin in rats	rh-insulin	Morimoto et al., 2001
HA-coated albumin nanoparticles for targeted peptide delivery to the retina	Connexin43 mimetic peptide	Huang et al., 2017
Chitosan-HA polyelectrolyte complex scaffold crosslinked with genipin for immobilization	Bone morphogenetic protein-2	Nath et al., 2015
HA hydrogels in a mouse subcutaneous environment	Bone morphogenetic protein-2	Todeschi et al., 2017
Proteolytically degradable HA hydrogels for bone repair	Bone morphogenetic protein-2	Holloway et al., 2015
pH-sensitive coiled-coil peptide-cross-linked ha nanogels for targeted intracellular protein delivery to cd44 positive cancer cells	Cytochrome C	Ding et al., 2018
Sodium hyaluronate-based dry powder formulations after pulmonary delivery to beagle dogs	Insulin	Surendrakumar et al., 2003
Polyion complex of chitosan and sodium hyaluronate as an implant device	Insulin	Surini, 2003
Tissue engineering scaffolds with a poly(lactic-co-glycolic acid) grafted HA conjugate encapsulating an intact bone morphogenetic protein-2/poly(ethylene glycol) complex	Bone morphogenetic protein-2	Park et al., 2011

1.5 CONCLUSION AND FUTURE PERSPECTIVE

The application of HA, as a biodegradable polymer for controlled-release and targeted drug delivery in pharmaceutical research, has invited a lot of interest. Still, the majority of the studies are inceptive and not supported even with *in-vivo* studies. However, it is still understood that HA has a good prospect of being used as a pharmaceutical additive such as drug carriers, as release retardant and many more in the formulation development process. Moreover, the application of HA for drug targeting looks promising, but research is still limited, and further studies are needed to establish it to be drug carrier at target tissues. Chemically modified HA also looks promising, but limited research data limits its usage in pharmaceutical applications. More safety data is required to expand the application of chemically modified HA for human use. The complex process of extraction and synthesis further limits its industrial application. The success of HA in medical and cosmetics science is surely enticing, but the industrial and clinical application is still a long way ahead.

KEYWORDS

- **Association of Pharmaceutical Teachers of India**
- **epidermal growth factor**
- **hyaluronan syntheses**
- **hyaluronic acid**
- **intramuscular**
- **intravenous**

REFERENCES

Adhikari, S. N. R., Nayak, B. S., Nayak, A. K., & Mohanty, B., (2010). Formulation and evaluation of buccal patches for delivery of atenolol. *AAPS Pharm. Sci. Tech., 11*, 1038–1044.

Ali, M., Singh, S., Kumar, A., Singh, S., Ansari, M., & Pattnaik, G., (2010). Preparation and *in vitro* evaluation of sustained release matrix tablets of phenytoin sodium using natural polymers. *Int. J. Pharm. Pharmaceut. Sci., 2*, 174–179.

Anirudhan, T. S., Nair, S. S., & Nair, A. S., (2016). Fabrication of a bioadhesive transdermal device from chitosan and hyaluronic acid for the controlled release of lidocaine. *Carbohydr. Polym.*, *152*, 687–698.

Aragona, P., (2002). Sodium hyaluronate eye drops of different osmolarity for the treatment of dry eye in Sjogren's syndrome patients. *British J. Ophthal.*, *86*, 879–884.

Baier, L. J., & Schmidt, C. E., (2004). *Hyaluronan*. In: *Encyclopedia of Biomaterials and Biomedical Engineering* (pp. 779–789). Taylor & Francis.

Banzato, A., Bobisse, S., Rondina, M., Renier, D., Bettella, F., Esposito, G., et al., (2008). A paclitaxel-hyaluronan bioconjugate targeting ovarian cancer affords a potent *in vivo* therapeutic activity. *Clin. Cancer Res.*, *14*, 3598–3606.

Battistini, F. D., Tártara, L. I., Boiero, C., Guzmán, M. L., Luciani-Giaccobbe, L. C., Palma, S. D., Allemandi, D. A., Manzo, R. H., & Olivera, M. E., (2017). The role of hyaluronan as a drug carrier to enhance the bioavailability of extended-release ophthalmic formulations. Hyaluronan-timolol ionic complexes as a model case. *Eur. J. Pharm. Sci.*, *105*, 188–194.

Behera, A. K., Nayak, A. K., Mohanty, B., & Barik, B. B., (2010). Formulation and optimization of losartan potassium tablets. *Int. J. Appl. Pharm., 2*, 15–19.

Bera, H., Boddupalli, S., & Nayak, A. K., (2015b). Mucoadhesive-floating zinc-pectinate–sterculia gum interpenetrating polymer network beads encapsulating ziprasidone HCl. *Carbohydr. Polym.*, *131*, 108–118.

Bera, H., Boddupalli, S., Nandikonda, S., Kumar, S., & Nayak, A. K., (2015a). Alginate gel-coated oil-entrapped alginate–tamarind gum–magnesium stearate buoyant beads of risperidone. *Int. J. Biol. Macromol.*, *78*, 102–111.

Bera, H., Kandukuri, S. G., Nayak, A. K., & Boddupalli, S., (2015c). Alginate-sterculia gum gel-coated oil-entrapped alginate beads for gastroretentive risperidone delivery. *Carbohydr. Polym.*, *120*, 74–84.

Bettin, P., Di Matteo, F., Rabiolo, A., Fiori, M., Ciampi, C., & Bandello, F., (2016). Deep sclerectomy with mitomycin C and injectable cross-linked hyaluronic acid implant. *J. Glaucoma, 25*, e625–e629.

Bigucci, F., Abruzzo, A., Saladini, B., Gallucci, M. C., Cerchiara, T., & Luppi, B., (2015). Development and characterization of chitosan/hyaluronan film for transdermal delivery of thiocolchicoside. *Carbohydr. Polym.*, *130*, 32–40.

Bongiovì, F., Fiorica, C., Palumbo, F. S., Di Prima, G., Giammona, G., & Pitarresi, G., (2018). Imatinib-loaded micelles of hyaluronic acid derivatives for the potential treatment of neovascular ocular diseases. *Molecular Pharm.*, *15*, 5031–5045.

Bot, P., Hoefer, I., Piek, J., & Pasterkamp, G., (2008). Hyaluronic acid: Targeting immune modulatory components of the extracellular matrix in atherosclerosis. *Curr. Med. Chem.*, *15*, 786–791.

Brown, M. A., (1995). *The Effects of a Novel Formulation of Cyclosporin on Antibody and Cell-Mediated Immune Reactions in the Pleural Cavity of Rats* (pp. 121–131). London: Royal Society of Medicine Press.

Brown, M. B., & Jones, S. A., (2005). Hyaluronic acid: A unique topical vehicle for the localized delivery of drugs to the skin. *J. Eur. Acad. Dermat. Vener.*, *19*, 308–318.

Brown, M. B., Hanpanitcharoen, M., & Martin, G. P., (2001). An *in vitro* investigation into the effect of glycosaminoglycans on the skin partitioning and deposition of NSAIDs. *Int. J. Pharm.*, *225*, 113–121.

Hyaluronic Acid (Hyaluronan): Pharmaceutical Applications　　　　21

Brown, M., Marriott, C., & Martin, G. P., (1995). A *Study of the Transdermal Drug Delivery Properties of Hyaluronan* (pp. 53–71). London: Royal Society of Medicine Press.

Calles, J. A., López-García, A., Vallés, E. M., Palma, S. D., & Diebold, Y., (2016). Preliminary characterization of dexamethasone-loaded cross-linked hyaluronic acid films for topical ocular therapy. *Int. J. Pharm.*, *509*, 237–243.

Camber, O., & Edman, P., (1989). Sodium hyaluronate as an ophthalmic vehicle: Some factors governing its effect on the ocular absorption of pilocarpine. *Curr. Eye Res.*, *8*, 563–567.

Camber, O., Edman, P., & Gurny, R., (1987). Influence of sodium hyaluronate on the meiotic effect of pilocarpine in rabbits. *Curr. Eye Res.*, *6*, 779–784.

Chen, B., Miller, R. J., & Dhal, P. K., (2014). Hyaluronic acid-based drug conjugates: State-of-the-art and perspectives. *J. Biomed. Nanotechnol.*, *10*, 4–16.

Cho, H. J., Yoon, H. Y., Koo, H., Ko, S. H., Shim, J. S., Lee, J. H., Kim, K., Chan, K. I., & Kim, D. D., (2011). Self-assembled nanoparticles based on hyaluronic acid-ceramide (HA-CE) and Pluronic® for tumor-targeted delivery of docetaxel. *Biomater.*, *32*, 7181–7190.

Choi, H., Jeena, M. T., Palanikumar, L., Jeong, Y., Park, S., Lee, E., & Ryu, J. H., (2016). The HA-incorporated nanostructure of a peptide–drug amphiphile for targeted anticancer drug delivery. *Chem. Commun.*, *52*, 5637–5640.

Cowman, M. K., & Matsuoka, S., (2005). Experimental approaches to hyaluronan structure. *Carbohydr. Res.*, *340*, 791–809.

Das, B., Dutta, S., Nayak, A. K., & Nanda, U., (2014). Zinc alginate-carboxymethyl cashew gum microbeads for prolonged drug release: Development and optimization. *Int. J. Biol. Macromol.*, *70*, 505–515.

Das, B., Nayak, A. K., & Nanda, U., (2013). Topical gels of lidocaine HCl using cashew gum and Carbopol 940: Preparation and *in vitro* skin permeation. *Int. J. Biol. Macromol.*, *62*, 514–517.

Das, B., Sen, S. O., Maji, R., Nayak, A. K., & Sen, K. K., (2017). Transferosomal gel for transdermal delivery of risperidone: Formulation optimization and *ex vivo* permeation. *J. Drug Deliv. Sci. Technol.*, *38*, 59–71.

Di Cicco, M., Alicandro, G., Claut, L., Cariani, L., Luca, N., Defilippi, G., Costantini, D., & Colombo, C., (2014). Efficacy and tolerability of a new nasal spray formulation containing hyaluronate and tobramycin in cystic fibrosis patients with bacterial rhinosinusitis. *J. Cystic Fibrosis*, *13*, 455–460.

Ding, L., Jiang, Y., Zhang, J., Klok, H. A., & Zhong, Z., (2018). pH-sensitive coiled-coil peptide-cross-linked hyaluronic acid nanogels: Synthesis and targeted intracellular protein delivery to CD44 positive cancer cells. *Biomacromol.*, *19*, 555–562.

Dionisio, M., & Grenha, A., (2012). Locust bean gum: Exploring its potential for biopharmaceutical applications. *J. Pharm. Bioallied. Sci.*, *4*, 175–185.

Drury, J. L., & Mooney, D. J., (2003). Hydrogels for tissue engineering: Scaffold design variables and applications. *Biomater.*, *24*, 4337–4351.

Esposito, E., Pecorelli, A., Sguizzato, M., Drechsler, M., Mariani, P., Carducci, F., Cervellati, F., Nastruzzi, C., Cortesi, R., & Valacchi, G., (2018). Production and characterization of nanoparticle based hyaluronate gel containing retinyl palmitate for wound healing. *Curr. Drug Deliv.*, *15*, 1172–1182.

Fan, X., Zhao, X., Qu, X., & Fang, J., (2015). pH-sensitive polymeric complex of cisplatin with hyaluronic acid exhibits tumor-targeted delivery and improved *in vivo* antitumor effect. *Int. J. Pharm.*, *496*, 644–653.

Fan, Y., Yao, J., Du, R., Hou, L., Zhou, J., Lu, Y., Meng, Q., & Zhang, Q., (2012). Ternary complexes with core-shell bilayer for double level targeted gene delivery: *In vitro* and *in vivo* evaluation. *Pharm. Res.*, *30*, 1215–1227.

Fiorica, C., Palumbo, F. S., Pitarresi, G., Bongiovì, F., & Giammona, G., (2017). Hyaluronic acid and beta-cyclodextrins films for the release of corneal epithelial cells and dexamethasone. *Carbohyd. Polym.*, *166*, 281–290.

Foox, M., & Zilberman, M., (2015). Drug delivery from gelatin-based systems. *Expt. Opin. Drug Deliv.*, *12*, 1547–1563.

Fu, C., Li, H., Li, N., Miao, X., Xie, M., Du, W., & Zhang, L. M., (2015). Conjugating an anticancer drug onto thiolated hyaluronic acid by acid liable hydrazone linkage for its gelation and dual stimuli-response release. *Carbohydr. Polym.*, *128*, 163–170.

Galer, C. E., Sano, D., Ghosh, S. C., Hah, J. H., Auzenne, E., Hamir, A. N., Myers, J. N., & Klostergaard, J., (2011). Hyaluronic acid-paclitaxel conjugate inhibits growth of human squamous cell carcinomas of the head and neck via a hyaluronic acid-mediated mechanism. *Oral Oncol.*, *47*, 1039–1047.

Gao, Y., Cheng, X., Wang, Z., Wang, J., Gao, T., Li, P., Kong, M., & Chen, X., (2014). Transdermal delivery of 10,11-methylenedioxy camptothecin by hyaluronic acid-based nanoemulsion for inhibition of keloid fibroblast. *Carbohydr. Polym.*, *112*, 376–386.

Garg, H. G., & Hales, C. A., (2004). *Chemistry and Biology of Hyaluronan* (pp. 271–279), Elsevier.

Girish, K. S., & Kemparaju, K., (2006). Inhibition of Naja naja venom hyaluronidase: Role in the management of poisonous bite. *Life Sci.*, *78*, 1433–1440.

Grimaudo, M. A., Nicoli, S., Santi, P., Concheiro, A., & Alvarez-Lorenzo, C., (2018). Cyclosporine-loaded cross-linked inserts of sodium hyaluronan and hydroxypropyl-β-cyclodextrin for ocular administration. *Carbohydr. Polym.*, *201*, 308–316.

Guru, P. R., Bera, H., Das, M., Hasnain, M. S., & Nayak, A. K., (2018). Aceclofenac-loaded *Plantago ovata* F. husk mucilage-Zn^{+2}-pectinate controlled-release matrices. *Starch – Stärke.*, *70*, 1700136.

Guru, P. R., Nayak, A. K., & Sahu, R. K., (2013). Oil-entrapped sterculia gum-alginate buoyant systems of aceclofenac: Development and *in vitro* evaluation. *Colloids Surf. B: Biointerf.*, *104*, 268–275.

Hahn, S. K., Kim, J. S., & Shimobouji, T., (2007). Injectable hyaluronic acid microhydrogels for controlled release formulation of erythropoietin. *J. Biomed. Mater. Res. Part A*, *80A*, 916–924.

Hargittai, M., & Hargittai, I. B., (2011). *History of hyaluronan science: From basic science to clinical applications.* Pubmatrix.

Hasnain, M. S., & Nayak, A. K., (2018a). Chitosan as responsive polymer for drug delivery applications. In: *Makhlouf, A. S. H., & Abu-Thabit, N. Y., (eds.), Stimuli-Responsive Polymeric Nanocarriers for Drug Delivery Applications* (Vol. 1, pp. 581–605). Types and triggers, Woodhead Publishing Series in Biomaterials, Elsevier Ltd.

Hasnain, M. S., & Nayak, A. K., (2018b). Alginate-inorganic composite particles as sustained drug delivery matrices. In: Inamuddin, A. A. M., & Mohammad, A., (eds.), *Applications of Nanocomposite Materials in Drug Delivery* (pp. 39–74). A volume in Woodhead Publishing Series in Biomaterials, Elsevier Inc.

Hasnain, M. S., Ahmad, S. A., Chaudhary, N., Hoda, M. N., & Nayak, A. K., (2019). Biodegradable polymer matrix nanocomposites for bone tissue engineering. In: Inamuddin, A.

A. M., & Mohammad, A., (eds.), *Applications of Nanocomposite Materials in Orthopedics* (pp. 1–37). A volume in Woodhead Publishing Series in Biomaterials, Elsevier Inc.

Hasnain, M. S., Nayak, A. K., Singh, M., Tabish, M., Ansari, M. T., & Ara, T. J., (2016). Alginate-based biopolymeric-nanobioceramic composite matrices for sustained drug release. *Int. J. Biol. Macromol., 83*, 71–77.

Hasnain, M. S., Nayak, A. K., Singh, R., & Ahmad, F., (2010). Emerging trends of natural-based polymeric systems for drug delivery in tissue engineering applications. *Sci. J. UBU., 1*, 1–13.

Hasnain, M. S., Rishishwar, P., & Ali, S., (2017a). Use of cashew bark exudate gum in the preparation of 4% lidocaine HCl topical gels. *Int. J. Pharm. Pharmaceut. Sci., 9*, 146.

Hasnain, M. S., Rishishwar, P., & Ali, S., (2017b). Floating-bioadhesive matrix tablets of hydralazine HCl made of cashew gum and HPMC K4M. *Int. J. Pharm. Pharmaceut. Sci., 9*, 124.

Hasnain, M. S., Rishishwar, P., Rishishwar, S., Ali, S., & Nayak, A. K., (2018a). Extraction and characterization of cashew tree (*Anacardium occidentale*) gum, use in aceclofenac dental pastes. *Int. J. Biol. Macromol., 116*, 1074–1081.

Hasnain, M. S., Rishishwar, P., Rishishwar, S., Ali, S., & Nayak, A. K., (2018b). Isolation and characterization of *Linum usitatisimum* polysaccharide to prepare mucoadhesive beads of diclofenac sodium. *Int. J. Biol. Macromol., 116*, 162–172.

Hirakura, T., Yasugi, K., Nemoto, T., Sato, M., Shimoboji, T., Aso, Y., Morimoto, N., & Akiyoshi, K., (2010). Hybrid hyaluronan hydrogel encapsulating nanogel as a protein nanocarrier: New system for sustained delivery of protein with a chaperone-like function. *J. Control. Release, 142*, 483–489.

Holloway, J. L., Ma, H., Rai, R., Hankenson, K. D., & Burdick, J. A., (2015). Synergistic effects of SDF-1α and BMP-2 delivery from proteolytically degradable hyaluronic acid hydrogels for bone repair. *Macromol. Biosci., 15*, 1218–1223.

Hou, L., Feng, Q., Wang, Y., Yang, X., Ren, J., Shi, Y., Shan, X., Yuan, Y., Wang, Y., & Zhang, Z., (2015). Multifunctional hyaluronic acid modified graphene oxide loaded with mitoxantrone for overcoming drug resistance in cancer. *Nanotechnol., 27*, 015701.

Huang, D., Chen, Y. S., & Rupenthal, I. D., (2017). Hyaluronic acid coated albumin nanoparticles for targeted peptide delivery to the retina. *Mol. Pharmaceutics, 14*, 533–545.

Huang, G., & Huang, H., (2018). Application of hyaluronic acid as carriers in drug delivery. *Drug Deliv., 25*, 766–772.

Jana, S., Ali, S. A., Nayak, A. K., Sen, K. K., & Basu, S. K., (2014a). Development of topical gel containing aceclofenac-crospovidone solid dispersion by "Quality by Design (QbD)" approach. *Chem. Eng. Res. Des., 92*, 2095–2105.

Jana, S., Das, A., Nayak, A. K., Sen, K. K., & Basu, S. K., (2013). Aceclofenac-loaded unsaturated esterified alginate/gellan gum microspheres: *In vitro* and *in vivo* assessment. *Int. J. Biol. Macromol., 57*, 129–137.

Jana, S., Gangopadhaya, A., Bhowmik, B. B., Nayak, A. K., & Mukhrjee, A., (2015b). Pharmacokinetic evaluation of testosterone-loaded nanocapsules in rats. *Int. J. Biol. Macromol., 72*, 28–30.

Jana, S., Manna, S., Nayak, A. K., Sen, K. K., & Basu, S. K., (2014b). Carbopol gel containing chitosan-egg albumin nanoparticles for transdermal aceclofenac delivery. *Colloids Surf. B: Biointerf., 114*, 36–44.

Jana, S., Samanta, A., Nayak, A. K., Sen, K. K., & Basu, S. K., (2015a). Novel alginate hydrogel core-shell systems for combination delivery of ranitidine HCl and aceclofenac. *Int. J. Biol. Macromol.*, *74*, 85–92.

Jarvis, B., & Figgitt, D. P., (2003). Topical 3% diclofenac in 2.5% hyaluronic acid gel. *American J. Clin. Dermatol.*, *4*, 203–213.

Jena, A. K., Nayak, A. K., De, A., Mitra, D., & Samanta, A., (2018). Development of lamivudine containing multiple emulsions stabilized by gum odina. *Future J. Pharm. Sci.*, *4*, 71–79.

Jeong, Y. I., Kim, S. T., Jin, S. G., Ryu, H. H., Jin, Y. H., Jung, T. Y., Kim, I. Y., & Jung, S., (2008). Cisplatin-incorporated hyaluronic acid nanoparticles based on ion-complex formation. *J. Pharm. Sci.*, *97*, 1268–1276.

Jiang, T., Zhang, Z., Zhang, Y., Lv, H., Zhou, J., Li, C., Hou, L., & Zhang, Q., (2012). Dual-functional liposomes based on pH-responsive cell-penetrating peptide and hyaluronic acid for tumor-targeted anticancer drug delivery. *Biomater.*, *33*, 9246–9258.

Kim, A. R., Lee, S. L., & Park, S. N., (2018). Properties and in vitro drug release of pH- and temperature-sensitive double cross-linked interpenetrating polymer network hydrogels based on hyaluronic acid/poly (N-isopropylacrylamide) for transdermal delivery of luteolin. *Int. J. Biol. Macromol.*, *118*, 731–740.

Kong, B. J., Kim, A., & Park, S. N., (2016). Properties and *in vitro* drug release of hyaluronic acid-hydroxyethyl cellulose hydrogels for transdermal delivery of isoliquiritigenin. *Carbohydr. Polym.*, *147*, 473–481.

Kuotsu, K., & Bandyopadhyay, A. K., (2007). Development of oxytocin nasal gel using natural mucoadhesive agent obtained from the fruits of *Dillenia indica*. L. *Sci. Asia.*, *33*, 57.

Kurisawa, M., Chung, J. E., Yang, Y. Y., Gao, S. J., & Uyama, H., (2005). Injectable biodegradable hydrogels composed of hyaluronic acid–tyramine conjugates for drug delivery and tissue engineering. *Chem. Commun.*, *43*, 12.

Kyyrönen, K., Hume, L., Benedetti, L., Urtti, A., Topp, E., & Stella, V., (1992). Methylprednisolone esters of hyaluronic acid in ophthalmic drug delivery: *In vitro* and *in vivo* release studies. *Int. J. Pharm.*, *80*, 161–169.

Lee, J. Y., & Spicer, A. P., (2000). Hyaluronan: A multifunctional, mega Dalton, stealth molecule. *Current Opin. Cell Biol.*, *12*, 581–586.

Lee, J., Yoo, K. C., Ko, J., Yoo, B., Shin, J., Lee, S. J., & Sohn, D., (2017). Hollow hyaluronic acid particles by competition between adhesive and cohesive properties of catechol for anticancer drug carrier. *Carbohydr. Polym.*, *164*, 309–316.

Li, C., Chen, R., Xu, M., Qiao, J., Yan, L., & Guo, X. D., (2018). Hyaluronic acid modified MPEG-b-PAE block copolymer aqueous micelles for efficient ophthalmic drug delivery of hydrophobic genistein. *Drug Deliv.*, *25*, 1258–1265.

Lim, S. T., Forbes, B., Berry, D. J., Martin, G. P., & Brown, M. B., (2002). *In vivo* evaluation of novel hyaluronan/chitosan microparticulate delivery systems for the nasal delivery of gentamicin in rabbits. *Int. J. Pharm.*, *231*, 73–82.

Lin, W., & Maibach, H., (1996). Percutaneous absorption of diclofenac in hyaluronic acid gel: *In vitro* study in human skin. *Round Table Series-Royal Society of Medicine*, *45*, 167–173.

Liu, D., Lian, Y., Fang, Q., Liu, L., Zhang, J., & Li, J., (2018). Hyaluronic-acid-modified lipid-polymer hybrid nanoparticles as an efficient ocular delivery platform for moxifloxacin hydrochloride. *Int. J. Biol. Macromol.*, *116*, 1026–1036.

Liu, E., Zhou, Y., Liu, Z., Li, J., Zhang, D., Chen, J., & Cai, Z., (2015). Cisplatin loaded hyaluronic acid modified TiO_2 nanoparticles for neoadjuvant chemotherapy of ovarian cancer. *J. Nanomater.*, 1–8.

Liu, S., Jin, M. N., Quan, Y. S., Kamiyama, F., Katsumi, H., Sakane, T., & Yamamoto, A., (2012). The development and characteristics of novel microneedle arrays fabricated from hyaluronic acid, and their application in the transdermal delivery of insulin. *J. Control. Release.*, *161*, 933–941.

Liu, Y., Sun, J., Cao, W., Yang, J., Lian, H., Li, X., Sun, Y., Wang, Y., Wang, S., & He, Z., (2011). Dual targeting folate-conjugated hyaluronic acid polymeric micelles for paclitaxel delivery. *Int. J. Pharm.*, *421*, 160–169.

Luo, Y., Wang, X., Du, D., & Lin, Y., (2015). Hyaluronic acid-conjugated apoferritin nanocages for lung cancer targeted drug delivery. *Biomater. Sci.*, *3*, 1386–1394.

Maji, R., Ray, S., Das, B., & Nayak, A. K., (2012). Ethyl cellulose microparticles containing metformin HCl by emulsification-solvent evaporation technique: Effect of formulation variables. *ISRN Polym. Sci.*, 1–7.

Malakar, J., & Nayak, A., (2013). Floating bioadhesive matrix tablets of ondansetron HCl: Optimization of hydrophilic polymer-blends. *Asian J. Pharm.*, *7*, 174.

Malakar, J., Basu, A., & Nayak, A. K., (2014a). Candesartan cilexetil microemulsions for transdermal delivery: Formulation, *in-vitro* skin permeation and stability assessment. *Curr. Drug Deliv.*, *11*, 313–321.

Malakar, J., Das, K., & Nayak, A. K., (2014). *In situ* cross-linked matrix tablets for sustained salbutamol sulfate release–Formulation development by statistical optimization. *Polim. Med.*, *44*, 221–230.

Malakar, J., Dutta, P., Purokayastha, S. D., Dey, S., & Nayak, A. K., (2014). Floating capsules containing alginate-based beads of salbutamol sulfate: *In vitro-in vivo* evaluations. *Int. J. Biol. Macromol.*, *64*, 181–189.

Malakar, J., Nayak, A. K., & Das, A., (2013a). Modified starch (cationized)-alginate beads containing aceclofenac: Formulation optimization using central composite design. *Starch – Stärke.*, *65*, 603–612.

Malakar, J., Nayak, A. K., & Goswami, S., (2012). Use of response surface methodology in the formulation and optimization of bisoprolol fumarate matrix tablets for sustained drug release. *ISRN Pharm.*, 1–10.

Mcewan, L. E., & Smith, J. G., (1997). Topical diclofenac/hyaluronic acid gel in the treatment of solar keratoses. *Austral. J. Dermatol.*, *38*, 187–189.

Menetrey, J., Zaffagnini, S., Fritschy, D., & Van Dijk, C. N., (2012). *ESSKA Instructional Course Lecture Book: Geneva 2012* (pp. 1–286). Springer Science & Business Media.

Meyer, K., & Palmer, J. W., (1936). On glycoproteins II. The polysaccharides of vitreous humor and of umbilical cord. *J. Biol. Chem.*, *114*, 689–703.

Mio, K., & Stern, R., (2002). Inhibitors of the hyaluronidases. *Matrix Biol.*, *21*, 31–37.

Moreira, C. A., Armstrong, D. K., Jelliffe, R. W., Moreira, A. T., Woodford, C. C., Liggett, P. E., & Trousdale, M. D., (1991a). Sodium hyaluronate as a carrier for intravitreal gentamicin an experimental study. *Acta Ophthalmol.*, *69*, 45–49.

Moreira, C. A., Moreira, A. T., Armstrong, D. K., Jelliffe, R. W., Woodford, C. C., Liggett, P. E., & Trousdale, M. D., (1991b). *In vitro* and *in vivo* studies with sodium hyaluronate as a carrier for intraocular gentamicin. *Acta Ophthalmol.*, *69*, 50–56.

Morimoto, K., Metsugi, K., Katsumata, H., Iwanaga, K., & Kakemi, M., (2001). Effects of low-viscosity sodium hyaluronate preparation on the pulmonary absorption of rh-insulin in rats. *Drug Dev. Ind. Pharm.*, *27*, 365–371.

Morimoto, K., Yamaguchi, H., Iwakura, Y., Morisaka, K., Ohashi, Y., & Nakai, Y., (1991). Effects of viscous hyaluronate-sodium solutions on the nasal absorption of vasopressin and an analog. *Pharm. Res.*, *8*, 471–474.

Motokawa, K., Hahn, S. K., Nakamura, T., Miyamoto, H., & Shimoboji, T., (2006). Selectively crosslinked hyaluronic acid hydrogels for sustained release formulation of erythropoietin. *J. Biomed. Mater. Res. Part A*, *78A*, 459–465.

Moustafa, M. A., Elnaggar, Y. S. R., El-Refaie, W. M., & Abdallah, O. Y., (2017). Hyalugel-integrated liposomes as a novel ocular nanosized delivery system of fluconazole with promising prolonged effect. *Int. J. Pharm.*, *534*, 14–24.

Nair, L. S., & Laurencin, C. T., (2006). Polymers as biomaterials for tissue engineering and controlled drug delivery. *Adv. Biochem. Eng. Biotechnol.*, *102*, 47–90.

Nath, S. D., Abueva, C., Kim, B., & Lee, B. T., (2015). Chitosan–hyaluronic acid polyelectrolyte complex scaffold crosslinked with genipin for immobilization and controlled release of BMP-2. *Carbohydr. Polym.*, *115*, 160–169.

Nayak, A. K., & Das, B., (2018). Introduction to polymeric gels. In: Pal, K., & Bannerjee, I., (eds.), *Polymeric Gels Characterization, Properties and Biomedical Applications* (pp. 3–27). A volume in Woodhead Publishing Series in Biomaterials, Elsevier Ltd.

Nayak, A. K., & Manna, K., (2011). Current developments in orally disintegrating tablet technology. *J. Pharm. Educ. Res.*, *2*, 24–38.

Nayak, A. K., & Pal, D. K., (2015a). Chitosan-based interpenetrating polymeric network systems for sustained drug release. In: Tiwari, A., Patra, H. K., & Choi, J. W., (eds.), *Advanced Theranostics Materials* (pp. 183–208). WILEY-Scrivener, U.S.A.

Nayak, A. K., & Pal, D., (2012). Natural polysaccharides for drug delivery in tissue engineering. *Everyman's Sci.*, *XLVI*, 347–352.

Nayak, A. K., & Pal, D., (2013a). Blends of jackfruit seed starch–pectin in the development of mucoadhesive beads containing metformin HCl. *Int. J. Biol. Macromol.*, *62*, 137–145.

Nayak, A. K., & Pal, D., (2013b). Ionotropically-gelled mucoadhesive beads for oral metformin HCl delivery: Formulation, optimization and antidiabetic evaluation. *J. Sci. Ind. Res.*, *72*, 15–22.

Nayak, A. K., & Pal, D., (2014). *Trigonella foenum-graecum* L. seed mucilage-gellan mucoadhesive beads for controlled release of metformin HCl. *Carbohydr. Polym.*, *107*, 31–40.

Nayak, A. K., & Pal, D., (2016a). Plant-derived polymers: Ionically gelled sustained drug release systems. In: Mishra, M., (ed.), *Encyclopedia of Biomedical Polymers and Polymeric Biomaterials* (Vol. 8, pp. 6002–6017). Taylor & Francis Group, New York, NY 10017, U.S.A.

Nayak, A. K., & Pal, D., (2016b). Sterculia gum-based hydrogels for drug delivery applications. In: Kalia, S., (ed.), *Polymeric Hydrogels as Smart Biomaterials, Springer Series on Polymer and Composite Materials* (pp. 105–151). Springer International Publishing, Switzerland.

Nayak, A. K., & Pal, D., (2017a). Tamarind seed polysaccharide: An emerging excipient for pharmaceutical use. *Indian J. Pharm. Educ. Res.*, *51*, S136–S146.

Nayak, A. K., & Pal, D., (2017c). Natural starches-blended ionotropically-gelled microparticles/beads for sustained drug release. In: Thakur, V. K., Thakur, M. K., & Kessler, M. R., (eds.), *Handbook of Composites from Renewable Materials* (Vol. 8, pp. 527–560). Nanocomposites: Advanced applications, WILEY-Scrivener, USA.

Nayak, A. K., & Pal, D., (2018). Functionalization of tamarind gum for drug delivery. In: Thakur, V. K., & Thakur, M. K., (eds.), *Functional Biopolymers* (pp. 35–56). Springer International Publishing, Switzerland.

Nayak, A. K., (2010). Advances in therapeutic protein production and delivery. *Int. J. Pharm. Pharmaceut. Sci.*, *2*, 1–5.

Nayak, A. K., (2016). Tamarind seed polysaccharide-based multiple-unit systems for sustained drug release. In: Kalia, S., & Averous, L., (eds.), *Biodegradable and Bio-Based Polymers: Environmental and Biomedical Applications* (pp. 471–494). WILEY-Scrivener, USA.

Nayak, A. K., Ahmad, S. A., Beg, S., Ara, T. J., & Hasnain, M. S., (2018a). Drug delivery: Present, past and future of medicine. In: Inamuddin, A. A. M., & Mohammad, A., (eds.), *Applications of Nanocomposite Materials in Drug Delivery* (pp. 255–282). A volume in Woodhead Publishing Series in Biomaterials, Elsevier Inc.

Nayak, A. K., Ara, T. J., Hasnain, M. S., & Hoda, N., (2018b). Okra gum-alginate composites for controlled releasing drug delivery. In: Inamuddin, A. A. M., & Mohammad, A., (eds.), *Applications of Nanocomposite Materials in Drug Delivery* (pp. 761–785). A volume in Woodhead Publishing Series in Biomaterials, Elsevier Inc.

Nayak, A. K., Bera, H., Hasnain, M. S., & Pal, D., (2018c). Graft-copolymerization of plant polysaccharides. In: Thakur, V. K., (ed.), *Biopolymer Grafting* (pp. 1–62). Synthesis and Properties, Elsevier Inc.

Nayak, A. K., Das, B., & Maji, R., (2012a). Calcium alginate/gum Arabic beads containing glibenclamide: Development and *in vitro* characterization. *Int. J. Biol. Macromol.*, *51*, 1070–1078.

Nayak, A. K., Das, B., & Maji, R., (2013a). Gastroretentive hydrodynamically balanced systems of ofloxacin: *In vitro* evaluation. *Saudi Pharm. J.*, *21*, 113–117.

Nayak, A. K., Hasnain, M. S., & Pal, D., (2018d). Gelled microparticles/beads of sterculia gum and tamarind gum for sustained drug release. In: Thakur, V. K., & Thakur, M. K., (eds.), *Polymeric Gel* (pp. 361–414). Springer International Publishing, Switzerland.

Nayak, A. K., Hasnain, M. S., Beg, S., & Alam, M. I., (2010a). Mucoadhesive beads of gliclazide: Design, development and evaluation. *Sci. Asia.*, *36*, 319–325.

Nayak, A. K., Maji, R., & Das, B., (2010). Gastroretentive drug delivery systems: A review. *Asian J. Pharm. Clin. Res.*, *3*, 2–10.

Nayak, A. K., Malakar, J., Pal, D., Hasnain, M. S., & Beg, S., (2018e). Soluble starch-blended Ca^{2+}-Zn^{2+}-alginate composites-based microparticles of aceclofenac: Formulation development and *in vitro* characterization. *Future J. Pharm. Sci.*, *4*, 63–70.

Nayak, A. K., Pal, D., & Das, S., (2013c). Calcium pectinate-fenugreek seed mucilage muco-adhesive beads for controlled delivery of metformin HCl. *Carbohydr. Polym.*, *96*, 349–357.

Nayak, A. K., Pal, D., & Hasnain, M. S., (2013b). Development and optimization of jackfruit seed starch-alginate beads containing pioglitazone. *Curr. Drug Deliv.*, *10*, 608–619.

Nayak, A. K., Pal, D., & Santra, K., (2013e). Plantago *ovata* F. Mucilage-alginate mucoadhesive beads for controlled release of glibenclamide: Development, optimization, and *in vitro-in vivo* evaluation. *J. Pharm.*, Article ID 151035.

Nayak, A. K., Pal, D., & Santra, K., (2014a). Ispaghula mucilage-gellan mucoadhesive beads of metformin HCl: Development by response surface methodology. *Carbohydr. Polym.*, *107*, 41–50.

Nayak, A. K., Pal, D., & Santra, K., (2014b). Development of calcium pectinate-tamarind seed polysaccharide mucoadhesive beads containing metformin HCl. *Carbohydr. Polym.*, *101*, 220–230.

Nayak, A. K., Pal, D., & Santra, K., (2014c). Tamarind seed polysaccharide–gellan mucoadhesive beads for controlled release of metformin HCl. *Carbohydr. Polym.*, *103*, 154–163.

Nayak, A. K., Pal, D., & Santra, K., (2014d). *Artocarpus heterophyllus* L. seed starch-blended gellan gum mucoadhesive beads of metformin HCl. *Int. J. Biol. Macromol.*, *65*, 329–339.

Nayak, A. K., Pal, D., & Santra, K., (2016). Swelling and drug release behavior of metformin HCl-loaded tamarind seed polysaccharide-alginate beads. *Int. J. Biol. Macromol.*, *82*, 1023–1027.

Nayak, A. K., Pal, D., Pany, D. R., & Mohanty, B., (2010b). Evaluation of *Spinacia oleracea* L. leaves mucilage as innovative suspending agent. *J. Adv. Pharm. Technol. Res.*, *1*, 338–341.

Nayak, A. K., Pal, D., Pradhan, J., & Ghorai, T., (2012c). The potential of *Trigonella foenum-graecum* L. seed mucilage as suspending agent. *Indian J. Pharm. Educ. Res.*, *46*, 312–317.

Nayak, A. K., Pal, D., Pradhan, J., & Hasnain, M. S., (2013d). Fenugreek seed mucilage-alginate mucoadhesive beads of metformin HCl: Design, optimization and evaluation. *Int. J. Biol. Macromol.*, *54*, 144–154.

Nayak, A., & Malakar, J., (2010). Formulation and *in vitro* evaluation of gastroretentive hydrodynamically balanced system for ciprofloxacin HCl. *J. Pharm. Educ. Res.*, *1*, 65–68.

Nayak, A., (2010). Advances in therapeutic protein production and delivery. *Int. J. Pharm. Pharmaceut. Sci.*, *2*, 1–5.

Nayak, A., Khatua, S., Hasnain, M. S., & Sen, K., (2011). Development of alginate-PVP K 30 microbeads for controlled diclofenac sodium delivery using central composite design. *DARU J. Pharm. Sci.*, *19*, 356–366.

Necas, J., Bartosikova, L., Brauner, P., & Kolar, J., (2008). Hyaluronic acid (hyaluronan): A review. *Veter. Med.*, *53*, 397–411.

Nguyen, V. D., Zheng, S., Han, J., Le, V. H., Park, J. O., & Park, S., (2017). Nanohybrid magnetic liposome functionalized with hyaluronic acid for enhanced cellular uptake and near-infrared-triggered drug release. *Colloids Surf. B: Biointerf.*, *154*, 104–114.

Nishinari, K., & Takahashi, R., (2003). Interaction in polysaccharide solutions and gels. *Curr. Opin. Colloid Interf. Sci.*, *8*, 396–400.

Ossipov, D. A., (2010). Nanostructured hyaluronic acid-based materials for active delivery to cancer. *Expert Opin. Drug Deliv.*, *7*, 681–703.

Pal, D., & Nayak, A. K., (2015a). Interpenetrating polymer networks (IPNs): Natural polymeric blends for drug delivery. In: Mishra, M., (ed.), *Encyclopedia of Biomedical Polymers and Polymeric Biomaterials* (Vol. 6, pp. 4120–4130). Taylor & Francis Group, New York, NY 10017, U.S.A.

Pal, D., & Nayak, A. K., (2015b). Alginates, blends and microspheres: Controlled drug delivery. In: Mishra, M., (ed.), *Encyclopedia of Biomedical Polymers and Polymeric Biomaterials* (Vol. I, pp. 89–98). Taylor & Francis Group, New York, NY 10017, U.S.A.

Pal, D., & Nayak, A. K., (2017). Plant polysaccharides-blended ionotropically-gelled alginate multiple-unit systems for sustained drug release. In: Thakur, V. K., Thakur, M. K., & Kessler, M. R., (eds.), *Handbook of Composites from Renewable Materials* (Vol. 6, pp. 399–400). Polymeric Composites, WILEY-Scrivener, U.S.A.

Pal, D., Nayak, A. K., & Saha, S., (2018). Interpenetrating polymer network hydrogels of chitosan: Applications in controlling drug release. In: *Mondal, I. H., (ed.), Cellulose-Based Superabsorbent Hydrogels, Polymers and Polymeric Composites: A Reference Series* (pp. 1–41). Springer, Cham.

Park, J. K., Shim, J. H., Kang, K. S., Yeom, J., Jung, H. S., Kim, J. Y., Lee, K. H., Kim, T. H., Kim, S. Y., Cho, D. W., & Hahn, S. K., (2011). Solid free-form fabrication of tissue-engineering scaffolds with a poly(lactic-co-glycolic acid) grafted hyaluronic acid conjugate encapsulating an intact bone morphogenetic protein-2/poly(ethylene glycol) complex. *Adv. Funct. Mater., 21*, 2906–2912.

Peer, D., & Margalit, R., (2003). Loading mitomycin C inside long circulating hyaluronan targeted nano-liposomes increases its antitumor activity in three mice tumor models. *Int. J. Cancer, 108*, 780–789.

Ray, S., Sinha, P., Laha, B., Maiti, S., Bhattacharyya, U. K., & Nayak, A. K., (2018). Polysorbate 80 coated crosslinked chitosan nanoparticles of ropinirole hydrochloride for brain targeting. *J. Drug Deliv. Sci. Technol., 48*, 21–29.

Rivers, J. K., (1997). An open study to assess the efficacy and safety of topical 3% diclofenac in a 2.5% hyaluronic acid gel for the treatment of actinic keratoses. *Arch. Dermatol., 133*, 1239.

Rivkin, I., Cohen, K., Koffler, J., Melikhov, D., Peer, D., & Margalit, R., (2010). Paclitaxel-clusters coated with hyaluronan as selective tumor-targeted nanovectors. *Biomater., 31*, 7106–7114.

Rolando, M., & Vagge, A., (2017). Safety and efficacy of cortisol phosphate in hyaluronic acid vehicle in the treatment of dry eye in Sjogren syndrome. *J. Ocular Pharmacol. Therapeut., 33*, 383–390.

Rosato, A., Banzato, A., De Luca, G., Renier, D., Bettella, F., Pagano, C., Esposito, G., Zanovello, P., & Bassi, P., (2006). HYTAD1-p20: A new paclitaxel-hyaluronic acid hydrosoluble bioconjugate for treatment of superficial bladder cancer. *Urol. Oncol.: Sem. Orig. Invest., 24*, 207–215.

Rosso, F., Quagliariello, V., Tortora, C., Di Lazzaro, A., Barbarisi, A., & Iaffaioli, R. V., (2013). Cross-linked hyaluronic acid sub-micron particles: *In vitro* and *in vivo* biodistribution study in cancer xenograft model. *J. Mater. Sci.: Mater. Med., 24*, 1473–1481.

Saettone, M. F., Giannaccini, B., Chetoni, P., Torracca, M. T., & Monti, D., (1991). Evaluation of high- and low-molecular-weight fractions of sodium hyaluronate and an ionic complex as adjuvants for topical ophthalmic vehicles containing pilocarpine. *Int. J. Pharm., 72*, 131–139.

Sakurai, K., Miyazaki, K., Kodera, Y., Nishimura, H., Shingu, M., & Inada, Y., (1997). Anti-inflammatory activity of superoxide dismutase conjugated with sodium hyaluronate. *Glycoconj. J., 14*, 723–728.

Saleem, M., Taher, M., Sanaullah, S., Najmuddin, M., Ali, J., Humaira, S., & Roshan, S., (2008). Formulation and evaluation of tramadol hydrochloride rectal suppositories. *Indian J. Pharm. Sci., 70*, 640.

Salzillo, R., Schiraldi, C., Corsuto, L., D'agostino, A., Filosa, R., De Rosa, M., & La Gatta, A., (2016). Optimization of hyaluronan-based eye drop formulations. *Carbohydr. Polym.*, *153*, 275–283.

Sant, S., Swati, S., Awadhesh, K., Sajid, M., Pattnaik, G., Tahir, M., & Farheen, S., (2011). Hydrophilic polymers as release modifiers for primaquine phosphate: Effect of polymeric dispersion. *ARS Pharmaceutica, 52*, 19–25.

Sinha, P., Ubaidulla, U., A. K., & Nayak, A. K., (2015a). Okra (*Hibiscus esculentus*) gum-alginate blend mucoadhesive beads for controlled glibenclamide release. *Int. J. Biol. Macromol., 72*, 1069–1075.

Sinha, P., Ubaidulla, U., Hasnain, M. S., Nayak, A. K., & Rama, B., (2015b). Alginate-okra gum blend beads of diclofenac sodium from aqueous template using $ZnSO_4$ as a cross-linker. *Int. J. Biol. Macromol., 79*, 555–563.

Sk, A., (2000). The effect of homogenized skin on the activity of licosamde antibiotics. *Proceedings of Millennium Congress of Pharmaceutical Science.* San Francisco, USA.

Song, E., Han, W., Li, C., Cheng, D., Li, L., Liu, L., Zhu, G., Song, Y., & Tan, W., (2014). Hyaluronic acid-decorated graphene oxide nanohybrids as nanocarriers for targeted and pH-responsive anticancer drug delivery. *ACS Appl. Mater. Interf., 6*, 11882–11890.

Soumya, R. S., Ghosh, S., & Abraham, E. T., (2010). Preparation and characterization of guar gum nanoparticles. *Int. J. Biol. Macromol., 46*, 267–269.

Sung, M. H., Park, C., Choi, J. C., Uyama, H., & Park, S. L., (2014). *Hyaluronidase Inhibitor Containing Poly-Gamma-Glutamic Acid as an Effective Component.* Google Patents.

Surendrakumar, K., Martyn, G. P., Hodgers, E. C. M., Jansen, M., & Blair, J. A., (2003). Sustained release of insulin from sodium hyaluronate-based dry powder formulations after pulmonary delivery to beagle dogs. *J. Control. Release, 91*, 385–394.

Surini, S., (2003). Polyion complex of chitosan and sodium hyaluronate as an implant device for nasal insulin delivery. *STP Pharma Sci., 13*, 265–275.

Tahir, M., Awadhesh, K., Swati, S., Sant, S., Sajid, M., & Pattnaik, G., (2010). Optimization of fast disintegrating tablets for diclofenac sodium using isabgol mucilage as super disintegrant. *Indian J. Pharm. Sci., 2*, 496–501.

Takayama, K., Hirata, M., Machida, Y., Masada, T., Sannan, T., & Nagai, T., (1990). Effect of interpolymer complex formation on bioadhesive property and drug release phenomenon of compressed tablet consisting of chitosan and sodium hyaluronate. *Chem. Pharm. Bull., 38*, 1993–1997.

Todeschi, M. R., El Backly, R. M., Varghese, O. P., Hilborn, J., Cancedda, R., & Mastrogiacomo, M., (2017). Host cell recruitment patterns by bone morphogenetic protein-2 releasing hyaluronic acid hydrogels in a mouse subcutaneous environment. *Regen. Med., 12*, 525–539.

Toole, B. P., (2000). Hyaluronan is not just a goo! *J. Clin. Invest., 106*, 335–336.

Verma, A., Dubey, J., Verma, N., & Nayak, A. K., (2017). Chitosan-hydroxypropyl methylcellulose matrices as carriers for hydrodynamically balanced capsules of moxifloxacin HCl. *Curr. Drug Deliv., 14*, 83–90.

Vishvkarma, A., Pattnaik, G., Ansari, M. T., & Ali, M. S., (2010). Pulmonary drug delivery system: A novel approach. *Inventi. Rapid: NDDS.*

Wang, L., & Jia, E., (2015). Ovarian cancer targeted hyaluronic acid-based nanoparticle system for paclitaxel delivery to overcome drug resistance. *Drug Deliv., 23*, 1810–1817.

Wei, S., Xie, J., Luo, Y., Ma, Y., Tang, S., Yue, P., & Yang, M., (2018). Hyaluronic acid based nanocrystals hydrogels for enhanced topical delivery of drug: A case study. *Carbohydr. Polym.*, *202*, 64–71.

Widjaja, L. K., Bora, M., Chan, P. N. P. H., Lipik, V., Wong, T. T. L., & Venkatraman, S. S., (2013). Hyaluronic acid-based nanocomposite hydrogels for ocular drug delivery applications. *J. Biomed. Mater. Res. Part A, 102*, 3056–3065.

Wolf, J. E., Taylor, J. R., Tschen, E., & Kang, S., (2001). Topical 3.0% diclofenac in 2.5% hyaluronan gel in the treatment of actinic keratoses. *Int. J. Dermatol.*, *40*, 709–713.

Xie, J., Ji, Y., Xue, W., Ma, D., & Hu, Y., (2018). Hyaluronic acid-containing ethosomes as a potential carrier for transdermal drug delivery. *Colloids Surf. B: Biointerf.*, *172*, 323–329.

Xin, D., Wang, Y., & Xiang, J., (2009). The use of amino acid linkers in the conjugation of paclitaxel with hyaluronic acid as drug delivery system: Synthesis, self-assembled property, drug release, and *in vitro* efficiency. *Pharm. Res.*, *27*, 380–389.

Xu, Y., Song, J., Pang, G., Chen, Z., Zhang, J., & Lü, X., (2004). Ocular pharmacokinetics of 0.5% pilocarpine with sodium hyaluronate in rabbits. *Chinese J. Ophthal.*, *40*, 87.

Yang, X., Dogan, I., Pannala, V. R., Kootala, S., Hilborn, J., & Ossipov, D., (2013). A hyaluronic acid–camptothecin nanoprodrug with cytosolic mode of activation for targeting cancer. *Polym. Chem.*, *4*, 4621.

Yerushalmi, N., Arad, A., & Margalit, R., (1994). Molecular and cellular studies of hyaluronic acid-modified liposomes as bioadhesive carriers for topical drug delivery in wound healing. *Arch. Biochem. Biophy.*, *313*, 267–273.

Yue, Y., Zhao, D., & Yin, Q., (2018). Hyaluronic acid modified nanostructured lipid carriers for transdermal bupivacaine delivery: *In vitro* and *in* vivo anesthesia evaluation. *Biomed. Pharmacother.*, *98*, 813–820.

Zeng, W., Li, Q., Wan, T., Liu, C., Pan, W., Wu, Z., Zhang, G., Pan, J., Qin, M., Lin, Y., Wu, C., & Xu, Y., (2016). Hyaluronic acid-coated niosomes facilitate tacrolimus ocular delivery: Mucoadhesion, precorneal retention, aqueous humor pharmacokinetics, and transcorneal permeability. *Colloids Surf. B: Biointerf.*, *141*, 28–35.

Zhu, M., Wang, J., & Li, N., (2018). A novel thermo-sensitive hydrogel-based on poly(N-isopropylacrylamide)/hyaluronic acid of ketoconazole for ophthalmic delivery. *Artif. Cells Nanomed. Biotechnol.*, *46*, 1282–1287.

Zhu, Z., Li, Y., Yang, X., Pan, W., & Pan, H., (2017). The reversion of anti-cancer drug antagonism of tamoxifen and docetaxel by the hyaluronic acid-decorated polymeric nanoparticles. *Pharmacol. Res.*, *126*, 84–96.

CHAPTER 2

Pharmaceutical Applications of Albumin

SUVADRA DAS[1] and PARTHA ROY[2]

[1]*Basic Science and Humanities Department,*
University of Engineering and Management, Kolkata, India

[2]*Department of Pharmaceutical Technology, Adamas University,*
Kolkata, India

ABSTRACT

Albumin is the most profuse blood protein. It serves as a reservoir and is also involved in the transport of a plethora of substances like nutrients, hormones, metals, and toxins. Albumin is produced mostly in liver hepatocytes. It has a molecular weight of 66.5 kDa, and its half-life is 19 days. It is. Albumin is available in three variants, i.e., ovalbumin, bovine serum albumin (BSA), and human serum albumin (HSA). Monomeric phosphoglycoprotein, ovalbumin finds immense application in food and drug delivery because of its easy availability, low cost, emulsion stabilization as well as pH/temperature responsive behavior. BSA is also extensively used in drug delivery because of its extraordinary ligand binding capacity. The most abundant hydrophilic plasma protein, HSA is another important component in different drug delivery system because of its inert and biodegradable nature with significant internalization in tumor and inflamed tissues.

These features of albumin make it an ideal candidate for half-life enhancement and site-directed delivery of bioactives associated by covalent conjugation, genetic fusions, or ligand-mediated interactions. The ability to the covalent or non-covalent attachment of the drug to albumin or encapsulation of drug into albumin nanoparticle depending on physical interaction gives a range of design option that has entered clinical trials or already in the market. This chapter will provide a spotlight on albumin structure and binding sites, a molecular association of albumin and the role of albumin in drug transport for the formulation design of next-generation drug delivery systems.

2.1 INTRODUCTION

The term albumin originated from the German word *albumen,* which means proteins. Albumin is one of the most investigated endogenous proteins. Natural biochemical and biophysical features of albumin make it an ideal candidate for the transportation of drugs. Albumin exists in three variants ovalbumin, bovine serum albumin (BSA), and human serum albumin (HSA). The multifunctional protein ovalbumin contains 385 amino acids with molecular weight 47,000 Da, and its isoelectric point is 4.8 (Oakenfull et al., 1997). Ovalbumin is extensively used in food (Elsadek et al., 2012) as well as in drug delivery because of certain advantages like low cost, easy availability, emulsion stabilization, and temperature/pH-responsive nature. BSA with molecular weight 69,323 Da and an isoelectric point 4.7 is also an inclusively used in drug delivery platform because of its low cost, abundance in nature, simplicity in puri-fication and excellent ligand binding capacity (Hu et al., 2006; Tantra et al., 2010). HSA is the most profuse multifunctional, non-glycosylated, anionic plasma protein with molecular weight 66,500 Da and means half-life of 19 days. HSA is stable in the pH range of 4–9 and when heated even at 60°C for 10 hours duration. HSA is biodegradable, non-toxic, and also lacks immunogenicity. The amino acid residues on albumin can be easily be associated with bioactives/targeting ligands through chemical conjugation, enhancing its value in the pharmaceutical application (Elsadek et al., 2012; Sleep et al., 2014).

Albumin is extensively synthesized by hepatocytes of circular polysomes on the rough endoplasmic reticulum. The average plasma concentration of albumin depends on its rate of synthesis, distribution of albumin between the intravascular and extravascular compartments, and degradation. Albumin constitutes approximately 60% of the total protein pool, i.e., about 3.5–5.0 g kg^{-1} body weight (250–300 g for a healthy 70 kg adult). But the plasma compartments contain only 42% of this pool, and the rest escapes into the extravascular compartments either via sinusitis and fenestrated capillaries or via an active transport mechanism into the continuous capillaries (Ganong, 1995). Some of the albumins are tissue bound and therefore not available freely in circulation. The small amounts of albumin (~2 g) are stored in the liver. Change in interstitial colloid osmotic pressure is the main regulatory factor (He et al., 1992) for synthesis of albumin other factors includes the nutritional state (supply of amino acid), energy supply (ATP and/or GTP), hormonal (insulin, glucagon, cortisol, thyroid) environment, and presence of systemic inflammatory response.

Albumin performs a number of important functions in blood like oncotic pressure maintenance, pH control, transport, and distribution of plethora of metabolic compound/endogenous ligands such as metals, fatty acids, hormones, amino acids, and toxins along with bioactive compounds by the help of its multiple binding sites and extended *in-vivo* half-life (~19 days) (Peters, 1996). Albumin is reported to have antioxidant property as well as enzymatic properties (Kragh-Hansen et al., 2002; Fasano et al., 2005).

2.1.1 ALBUMIN TYPES, STRUCTURE, AND BINDING SITES

HSA is a highly soluble, non-glycosylated globular protein (66kd) having 585 amino acid residues. It is characterized by low tryptophan residues but with loads of cysteine and charged amino acid residues like lysine, aspartic acids, glutamic acids (He et al., 1992). X-ray crystallographic examinations show that the polypeptide chain forms a heart shape tertiary structure with three homologous domains I, II, and III. Each domain contains two subdomains (A and B), which contains 4 and 6 α- helices, respectively (Figure 2.1). The subdomains move relative to one another by means of 9 flexible loops fastened together by 17 disulfide linkages. As a consequence, HSA is very stable to changes in pH, exposure to heat, and denaturing solvents. Subdomains with separate helical structure intervene the HSA binding with various endogenous and exogenous ligands with different ligand binding affinities and functions. The two main drug binding sites of HSA are Sundlow site I and Sundlow site II (Sudlow et al., 1975). Sundlow site I is located in subdomain IIA and preferentially binds with bulky heterocyclic compounds like warfarin (Kragh-Hansen et al., 2002; Petitpas et al., 2001). Sundlow site II is situated in subdomain IIIA and is termed as the benzodiazepine binding site with selective truss affinity for an aromatic compound like ibuprofen (Kragh-Hansen et al., 2002).

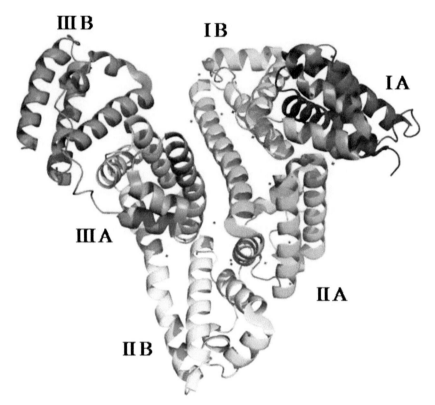

FIGURE 2.1 (See color insert.) The crystal structure of human serum albumin in complex with stearic acid. The three domains of albumin are shown in deep blue (IA), light blue (IB), green (IIA), yellow (IIB), light orange (IIIA), and deep orange (IIIB) (PDB 1e7e).

HSA contains 35 cysteine residues, and all of these except one (Cys34) are involved in disulfide bond formation that helps to stabilize the protein. Cys 34 is redox active for covalent attachment of drugs, and the thiol (-SH) group accounts for 80% of the thiols in plasma (Quinlan et al., 2005).

A large number of charged amino acid residues (83 basic and 98 acidic) in HSA decides its overall net charge and isoelectric point. At neutral pH, the calculated net charge of HSA decreases gradually from −9, −8 and +2 in domains I, II, and III respectively. This makes the overall net charge of −15 and isoelectric point ~4.8 (Demuro et al., 2005).

Drug and drug metabolites can also form covalent bonding with albumin. Acid glucuronide metabolite can form a covalent bond to HSA through a nucleophilic attack from NH_2, OH or SH faction of protein to the acyl carbon of the glucuronide without retention of the glucuronide moiety (Benet et al., 1993). Another mechanism describes the tautomerism of the sugar ring, which involves the migration of the acyl group from position 1 in the sugar ring to 2, 3, or 4 positions. Lysine group in protein reacts with an aldehyde in the open tautomer structure resulting in a covalent bond between drug and protein separated by glucuronic acid (Kragh-Hansen et al., 2002; Benet et al., 1993; Williams et al., 1994). Products of drug metabolism like furosemide, salicylic acid, or Nonsteroidal Anti-Inflammatory Drugs (NSAIDs) like ibuprofen, exhibit covalent association with HSA (Kragh-Hansen et al., 2002). Biocompatible and biodegradable HSA with minimal immunogenicity/ toxicity has been reported to exhibit selective internalization in tumor tissues making it the perfect device for drug and gene delivery (Elzoghby et al., 2012; Zheng et al., 2014).

2.1.2 BSA

Like HSA, BSA (MW ~ 66kDa) is a 585-residue, multi-domain, α-helical (no β sheet is there) protein that resembles a heart-like structure at physiological pH. Bovine (BSA) analog of HSA shares 76% sequence homology with HSA except for one more tryptophan residue in domain IA at the 134th position (in addition to the one present in domain IIA at position 212). The structural organization and compactness of BSA depends on pH and undergoes reversible conformational isomerization as a function of pH (Foster, 1977). Five conformational isomers are reported for BSA over different pH ranges namely, the E- form (Extended; pH<3), the F-form (Fast; pH 3–5), the N-form (Native; pH 5–7), the B-form (Basic; pH 7–8.5) and the A-form (Aged; pH>8.5). The interior of the protein is mostly hydrophobic, and both charged as well as apolar patches reside in the interface which gives the net charge –18 and isoelectric point of 4.7.

2.1.3 OVA

Ovalbumin is a monomeric phosphoglycoprotein with a molecular weight range between 42 and 47 kDa and isoelectric point around pH 4.8 (Oakenfull et al., 1997). The 3D structure of this multifunctional protein contains α- helical reactive loop, which is 41% of the molecule, 34% β-sheet, 12% β-turns and 13% random coils (Ngarize et al., 2004). Its chemical structure contains a disulfide bond and four free sulfhydryl groups. OVA is easily identified by the immune system compared to mammalian serum albumins and therefore often explored in vaccine delivery systems.

2.2 ALBUMIN RECEPTORS

The interaction with the cell surface receptor is responsible for degradation, cellular transcytosis, salvage, and recycling of albumin. Receptors include glycoproteins Gp18, Gp30 and Gp 60, a secreted protein, acidic, and rich in cysteine (SPARC), the Megalin/Cubilin complex and the neonatal Fc receptor (FcRn).

2.2.1 GLYCOPROTEIN RECEPTORS

The Gp18 and Gp30 are cell surface glycoprotein with molecular weight 18 and 30 kDa, respectively. Gp18 and Gp30 are located mainly in liver endothelium cell membranes (Ottnad et al., 1992) and peritoneal macrophages (Peters, 1996; Zhang et al., 1993). Gp18 and Gp30 resemble scavenger receptors as because they have a higher affinity for conformationally modified albumin than native one and intervene in the endocytosis and degradation of damaged albumin. Modification of native albumin occurs because of protection, normal aging, or disease mediated reaction. The altered albumin then binds with Gp18 and Gp 30 receptors either for internalization or degradation (Schnitzer et al., 1993; Schnitzer et al., 1993). Chemically modified BSA shows a 1000-fold higher affinity for Gp18 and Gp30 compared to native BSA (Schnitzer et al., 1992).

Gp 60 with molecular weight 60 kDa, also known as albondin and its presence is limited in vascular continuous endothelium and alveolar epithelium. Gp 60 binds native albumin and plays a pivotal role in the transcytosis of albumin and the bulk flow of plasma proteins across the intact endothelium from the apical to the basal membrane (Schnitzer et al., 1992, 1994; Tiruppathi et al., 1996).

Attachment of albumin to Gp60 in the endothelial cell surface results in accumulation of Gp60-albumin at the cell surface and alliance of caveolin-1 to form caveolae vesicle (Minshall et al., 2002; Malik et al., 2009). The caveolae then traverse through the cytoplasm, fuses with the basolateral membrane and release material from the caveolae into the interstitium.

Pharmaceutical Applications of Albumin   39

Approximately 50% of albumin transport is achieved by binding with Gp60 receptors while the rest occurs through fluid phase transport or through intercellular junctions (Schnitzer et al., 1993, 1994; Merlot et al., 2014).

2.2.2 SECRETED PROTEIN, ACIDIC, AND RICH IN CYSTEINE (SPARC) RECEPTOR

SPARC is a 43 kDa glycoprotein situated in the extracellular matrix (ECM). SPARC is linked with tissue growth, cell movement and/or cell proliferation (Sage et al., 1989; Brekken et al., 2000; Jacob et al., 1999; Kato et al., 1998; Lane et al., 1994; Pichler et al., 1996a, b) and is found in a variety of cells including fibroblasts and endothelial cells. It is overexpressed in cancer tissues and is reported to be associated with transport and accumulation of albumin at the tumor and inflamed sites (Said et al., 2013; Nagaraju et al., 2014; Podhajcer et al., 2008).

2.2.3 MEGALIN/ CUBILIN RECEPTOR

Cubilin is a 460 kDa glycoprotein found in the apical endocytic compartments of kidney proximal tubules, anchored to the membrane at the N-terminal by an α-helix. In spite of its large size, it is devoid of the transmembrane segment and cytoplasmic domain and therefore depends on another membrane protein, Megalin for endocytosis. Megalin is a 600 kDa large cellular glycoprotein. Megalin has an extracellular domain, a transmembrane segment as well as a cytoplasmic tail. Both cubilin and megalin are endocytic receptors, but cubilin fails to express any signal for endocytosis and interact with megalin to facilitate co-internalization. Both the receptors are primarily located in polarized epithelial cells with a few exceptions. Albumin attaches to Cubilin and Megalin and Megalin/ Cubilin complex to assist in receptor-mediated endocytosis and also to save albumin from excretion by kidneys (Birn et al., 2000; Cui et al., 1996).

2.2.4 NEONATAL FC RECEPTOR (FCRN)

The neonatal Fc receptor (FcRn) finds occurrence in different tissues and cells, including vascular, renal (podocytes and proximal convoluted tubule) and brain endothelial gut, upper airway, and alveolar epithelia. FcRn expression differs between animal species. It is expressed by human intestinal epithelial cells in both the neonate and adult but in case of rodent FcRn expression is highest in neonatal intestinal epithelial cells, but levels decline fast after weaning. FcRn is a heterodimeric type I glycoprotein comprising of an MHC-class I-like heavy chain (HC) and is composed of three extracellular domains (α1, α2, and α3), a single transmembrane domain and a 44 amino acid cytoplasmic tail. It is 45 kDa in humans and 51 kDa in rodents.

The neonatal Fc receptor (FcRn) regulates the serum half-lives of both IgG and albumin by a mechanism of increased association at acidic pH (<6.5) within the endosomes and further recycling and release into the extracellular space at physiological pH (Chaudhury et al., 2003; Ghetie et al., 1996; Junghans et al., 1996; Ward et al., 2009). Basis of the pH-dependent albumin association to FcRn have been elucidated by the Molecular modeling, X-ray crystallography and site-directed mutagenesis studies (Andersen et al., 2012; Schmidt et al., 2013; Sand et al., 2014; Oganesyan et al., 2014). Domain III of albumin was crucial for pH-dependent binding to FcRn. Domain I modulates and stabilizes binding, which is required for optimal interaction, and Domain II does not contribute to the interaction. Albumin binding to FcRn at a crevice between DI and DIII results in movement of both DI and DIII relative to the rest of the molecule, but the movement of DIII's is being the most significant. Studies of co-crystals of albumin: FcRn indicates that interaction of DIII with FcRn is predominantly hydrophobic in nature and centers around W53 and W59 within the FcRn HC (Schmidt et al., 2013). Albumin binding, like IgG, depends on the presence of conserved histidine residues (His166) in the α3 domain of FcRn discrete from that of IgG and three conserved histidine residues (H464, H510 and H535) on domain III of albumin (Andersen et al., 2012; Kenanova et al., 2010).

2.3 DRUG ASSOCIATION WITH ALBUMIN

The long circulatory half-life of albumin is because of the size of the albumin, which is greater than the renal threshold, and it undergoes specific interaction and recycling by FcRn. Small therapeutic peptides, proteins, and chemical drugs often suffer from poor therapeutic efficacy due to rapid clearance via the kidneys and liver. Their pharmacokinetic

Pharmaceutical Applications of Albumin 41

profile and physical characteristics can be improved by association with circulating endogenous albumin through non-covalent (by specific binding to albumin) or covalent (by conjugation and by direct genetic fusion) conjugation.

2.3.1 NON-COVALENT ASSOCIATION OF THERAPEUTIC MOLECULE TO ALBUMIN

Albumin acts as a transporter for a broad range of endogenous substances like fatty acid, bilirubin, and exogenous ligands like penicillins, warfarin through reversible non-covalent association. Non-covalent Van Der Waals force or electronic interactions are responsible for this association. The drug of interest conjugates with a molecule which has a binding affinity for albumin. For examples, fatty acids can be conjugated to the drug of interest, which upon injection associates with albumin. This strategy has been specifically exploited in the development of Levemir® (Insulin detemir) for the treatment of diabetes. Levemir® contains an insulin analog with an exposed lysine residue that is conjugated to myristic acid (Klein et al., 2007; Hermansen et al., 2006; Home et al., 2006). Victoza® (Lira-glutide) by Novo Nordisk is another diabetic drug which is a derivative of glucagon-like peptide-1 agonist (GLP-1) conjugated with a myristic acid at the ε-amino position of the N-terminal lysine. The drug after parenteral administration undergoes association with endogenous albumin in the blood through its fatty acid binding sites. Then drug slowly comes out of the association leading to enhancement of its half-life and bioavailability. Saturation of albumin with long Fatty Acid abolishes binding to FcRn receptor (Schmidt et al., 2013) which may be the limitations regarding half-life extension. Another example is "albutag" consisting of a small organic molecule, 4-(p-iodophenyl) butanoic acid derivatives that special affinity for albumin. The attachment of this tag to small molecules or recombinant proteins has been shown to be able to increase their half-lives. The tag has been conjugated to a free cysteine at the C-terminal end of a single-chain antibody (scFv) fragment with tumor specificity. That prolonged the half-life from ~20 min to 16.6 hours, in addition to improving tumor internalization (Trüssel et al., 2009).

Another approach is to use albumin binding antibody. Camelid antibodies constructed of heavy-chain antibody fragments containing a single variable domain (VHH) and two constant domains (CH_2 and CH_3). The

VNH domain is highly targeted specific and binds with their antigen target with high specificity and low viscosity in comparison to single chain antibodies. One example of camelid is murine trivalent antibodies that were constructed to bind HSA on the one hand and the murine or human tumor necrosis factor-α (TNF-α) on the other.

Camelid-derived albumin-binding nanobodies have been explored for tumor targeting. Peptide phage display technology has been used to develop peptides that have a high specific affinity for albumin. A panel of albumin binding peptides was fused to a Fab molecule with a specific affinity for human epidermal growth factor receptor 2 (EGFR2) (AB.Fab). To estimate the pharmacokinetic properties of an AB.Fab in humans, AB.Fab variants with similar affinities for rat and rabbit albumin were injected into rodents. It was demonstrated that AB.Fab has a direct correlation between albumin affinity and the desired pharmacokinetic profile (Nguyen et al., 2006).

The half-life of albumin among various animal species has been taken together to predict the terminal half-life and clearance of AB.Fab in human. The specific binding to albumin is retained when the albumin binding domain has attached the payload to increase the half-life of the payload. But this half-life extension is affected by the interaction of other molecules. The fusion of interleukin-1 receptor antagonist (IL-1ra) with an AlbudAb increased the half-life of IL-1ra from 2 min to 4.3 h, but the half-life recorded for the isolated albumin binding AlbudAb was 24 h (Holt et al., 2008).

Nanobody technology using noncovalent interaction with albumin has been explored for improvement of pharmacokinetic properties and tumor targeting. An example is bivalent anti-epidermal growth factor receptor (EGFR) nanobody (α-EGFR-α-EGFR) that binds to serum albumin (α-Alb). The α-EGFR-α-EGFR-α-Alb nanobody (50 kDa) was evaluated for biodistribution and tumor uptake and penetration in comparison to intact 150-kDa anti-EGFR mAb cetuximab in A431 xenograft-bearing nude mice. The nanobody and cetuximab was radiolabeled with ^{177}Lu to facilitate accurate quantitative analysis. Rapid blood clearance of ^{177}Lu-α-EGFR-α-EGFR causes the tumor to blood ratio > 80 within 6 h after injection. Tumor uptake was even decreased from 5.0 ± 1.4 to 1.1 ± 0.1% ID/g between 6 and 72 h after injection. Introduction of α-Alb increases the concentration of EGFR-α- EGFR-α-α-Alb in blood 21.2 ± 2.5, 11.9 ± 0.6, and 4.0 ± 1.4 ID/g and tumor levels are 19.4 ± 5.5, 35.2 ± 7.5, and 28.0 ± 6.8% ID/g at 6, 24, and 72 h after injection, respectively. EGFR-α-EGFR-α-α-Alb showed faster and deeper tumor penetration than cetuximab (Tijink et al., 2008).

2.3.2 COVALENT ASSOCIATION OF DRUG MOLECULE TO ALBUMIN

Another approach of utilizing albumin as a drug carrier is based on the covalent association of drugs to albumin by conjugation either through a chemical link or gene fusion. Small molecules like doxorubicin or methotrexate are conjugated with albumin in the development of anticancer, autoimmune, and anti-rheumatic therapeutics. Random conjugation of drugs to surface exposed multiple lysine residues on albumin resulted in a non-specific modification which interferes with binding to albumin receptors and increases clearance. Specific conjugation can be accomplished by conjugation to a single free thiol group on the cysteine residue at position 34. The Drug Affinity Complex (DAC) technology has been developed where peptides are modified to allow site-specific conjugation to either exogenous or endogenous albumin upon injection (Kratz et al., 2000). Technological advancements have led to conjugate formation prior to parenteral injection with recombinant human albumin (Preformed Conjugate-Drug Affinity Complex (PC-DAC)). This technology has been successfully applied to Exendin-4, a GLP-1 homologue (CJC-1131) for treatment of type 2 diabetes (Baggio et al., 2008; Kim et al., 2003; Léger et al., 2004) where the half-life has been increased from only a few hours for the GLP-1 analog to 9–15 days for the C34-conjugated drug (Giannoukakis, 2003). An acidic sensitive prodrug of doxorubicin is conjugated *in vivo* to C34 after intravenous (IV) administration for the treatment of sarcoma. The hydrazine linker connecting albumin and doxorubicin breaks upon exposure to the acidic environment in tumor tissues (Chawla et al., 2015).

Albumin fusion technology yields albumin protein conjugates by fusing the gene for human albumin to the gene that encodes the active protein drug. This technology has been applied to peptides, cytokines, coagulation factors, enzymes, hormones, growth factors, a variety of antibody fragments and redox regulators (Halpern et al., 2002; Wang et al., 2004; Melder et al., 2005; Duttaroy et al., 2005; Müller et al., 2007; Evans et al., 2010).

This technology increases the molecular weight of the native protein, thus prolongs the half-life in vivo. Recombinant interferon α2b was used for the treatment of chronic hepatitis C with a half-life of 4 hours in humans; however, upon genetic fusion to human albumin, the half-life increases to 141 hours (Bain et al., 2006). Masking of the protein molecule by albumin makes the protein molecule more resistant to proteases and less immunogenic. Another beneficial feature of genetic fusion is that as the therapeutic is synthesized as one transcript so further in vitro processing is not needed. A diverse variety of proteins that have been genetically glued to albumin requires different expression platforms depending upon the nature of fused protein. Some simple molecules like GLP-1, interferon α-2b, G-CSF, and antibody fragments can be produced from *Saccharomyces cerevisiae* and *Pichia pastoris*, but more complex molecules require specific post-translational modifications.

The circulatory half-life of human GLP-1 is normally 1–2 min (Halpern et al., 2002; Ahrén et al., 2009). The GLP-1 fusion, albiglutide™, is a protease-resistant GLP-1 receptor agonist made up of two repeats of the protease-resistant derivative of GLP-1 directly fused to the N-terminus of recombinant human albumin. Series of clinical trials with multiple biweekly or once-weekly injections of albiglutide shows safety, tolerability, and increased half-life (~5 days) (Madsbad et al., 2011). The G-CSF-fusion, balugrastim is comprised of granulocyte colony-stimulating factor protein directly linked to the C-terminus of albumin. Pre-clinical studies in mice and cynomolgus monkeys also showed the efficacy, well tolerability and increased the half-life of balugrastim in comparison to the unfused molecule (Halpern et al., 2002; Bock et al., 2010; Volovat et al., 2014).

2.4 ALBUMIN IN DRUG TRANSPORT

Albumin has diverse biological effects which include antioxidant and free radical scavenging effect, control of colloid osmotic pressure, acid-base balance and nitric oxide, association, and carrying of numerous elements like vitamins, hormones, electrolytes, enzymes, etc. within the blood and anticoagulant effect. It also helps in the modulation of vascular permeability by exerting antioxidant, anti-inflammatory, or anti-apoptotic effects (Vincent et al., 2014). Clinical use of albumin is dated back from World War II and is now used for different clinical conditions like cancer, sepsis, burns, hemorrhage, and in case of chronic renal and hepatic failure (Rozga et al., 2013). Among the different physiological effects of albumin, the most investigated pharmaceutical application of albumin is the transportation of various bioactive molecules. A PubMed search on the term "albumin nanoparticles" displayed a steep rise in the number of publications from 1996 to 2018. The product that cemented the use of albumin as a carrier is nanoparticle albumin-bound Paclitaxel which crossed the boundaries of laboratory examinations and clinical trials and reached the market. Albumin nanoparticle alone or in combination with drugs was subjected to routine clinical investigations in different advanced cancers. The results of the study exhibited sufficient reduction of toxicity in breast cancer patients treated with albumin nanoparticles with paclitaxel load compared to free paclitaxel (Gradishar et al., 2005). But when used in combination with drugs like carboplatin in extensive-stage small cell lung cancer, the results warranted frequent dose adjustments (Fu et al., 2011) (Figure 2.2).

Albumin-stabilized nanoparticle formulation of paclitaxel showed enhancement of maximum-tolerated dose, antitumor response, and life-span in mice xenograft models compared with solvent-based taxane formulations (Chen et al., 2015). The nanoparticles were also found to localize more in xenograft tumors and influence improved cytotoxicity than conventional Cremophor EL-paclitaxel (Chen et al., 2015). In Phase III clinical trial study, nanoparticle albumin-bound paclitaxel in combination with gemcitabine slightly enhanced the life span of pancreatic cancer patients (Hoffman et al., 2015). Similarly, results were observed in the case of combination therapy of curcumin and paclitaxel bound albumin nanoparticle, where the drug cocktail proved more cytotoxic than single-agent therapy (Ruttala et al., 2015). A cost-utility analysis study between paclitaxel bound nanoparticle albumin vs. castor oil-based standard paclitaxel formulation conducted on pretreated metastatic breast cancer revealed that the nanoparticle treatment was more economical chemo-therapeutic among the two (Alba et al., 2013). Experiments conducted to evaluate the influence of age of metastatic breast cancer patients on the pharmacokinetics and pharmacodynamics of albumin nanoparticle bound paclitaxel. Results showed that age had no significant influence on the pharmacodynamic response of the nano-formulation, but the significant relationship was observed between age and pharmacokinetic parameter 24-hour Area under the curve value (Hurria et al., 2015). Combination therapy of albumin-bound paclitaxel, doxorubicin, and cyclophosphamide in breast cancer patients containing a subset of triple negative breast cancer population achieved a pathological complete response following the chemotherapy regimen (Werner et al., 2017).

Kushwah et al., formulated gemcitabine conjugated albumin nanoparticles for the management of pancreatic cancer (Kushwah et al., 2017). The nanoparticles induced significant DNA damage and apoptosis compared to

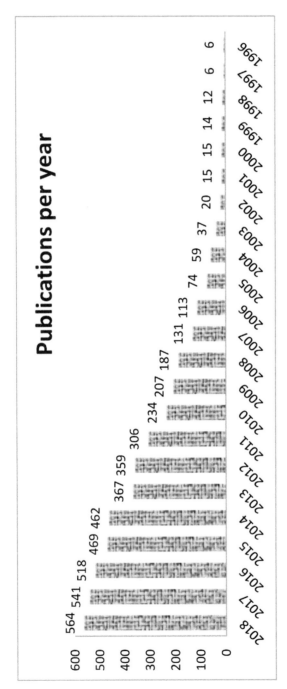

FIGURE 2.2 (See color insert.) Year-wise publications as PubMed search on the term "albumin nanoparticles."

free gemcitabine. Albumin nanoparticles with gabapentin load increased the concentration of the drug in the brain. Nanogabapentin significantly reduced the duration of all convulsion phases in both maximal electroshock and pentylenetetrazole stimulated experimental convulsion models compared to free gabapentin (Wilson et al., 2014). HSA nanoparticles were excellent solubility enhancers of hydrophobic bioactives like curcumin and also provided site-specific localization in the cytoplasm of tumor cells (Gong et al., 2015). Cabazitaxel encapsulated in albumin nanoparticle showed better uptake in tumors and extended blood circulation in prostate cancer xenograft models in nude mice (Qu et al., 2016). The nanoparticle associated toxicity, as evident from histopathological assays and blood levels of urea nitrogen and serum creatinine, were markedly less compared to free cabazitaxel treatment. BSA is hydrophilic in nature and can entrap different drug cargo based electrostatic interactions. Anti-inflammatory drug salicylic acid-loaded BSA nanoparticles exhibited a biphasic and pH-responsive drug release suitable for pharmaceutical applications (Bronze-Uhle et al., 2017). Such carriers are promising candidates where the rapid *in-vivo* release of salicylic acid is desired. In another study, acyclovir-loaded albumin nanoparticles were found to permeate easily through the human corneal epithelial T cells (Suwannoi et al., 2017) and therefore, can be explored as an ocular drug delivery system. Release profiles of glutaraldehyde cross-linked albumin nanoparticles loaded with another antiviral agent ganciclovir (Merodio et al., 2001) showed a sustained nature suitable for prolonging ocular residence as desired in retinal infections like cytomegalovirus retinitis. Drug delivery to the back of the eye is a serious challenge due to bioavailability issues. Albumin nanoparticles with a high load of aspirin and particle size of < 200nm produced greater than 10% discharge of drug cargo in the posterior chamber over a period of 3 days. Free aspirin-induced intense hemolysis in the ocular environment was overcome by its nanoparticulate formulation (Das et al., 2012). Stability studies of the formulation revealed no appreciable change during the study period of six months, which could affect its pharmacodynamic response.

Guo et al., reported that the caspase-independent pathway involving mitochondrial proteins also play a role in apoptosis of human ovarian cancer SKOV3 cells by resveratrol albumin nanoparticles (Guo et al., 2015). Particle engineering of HSA nanoparticles loaded with resveratrol produced site-specific and long circulating nanoparticles which were significantly toxic to PANC-1 cells compared to free the plant bioactive. Alongside *in-vivo* evaluations revealed that particle localization in tumor tissues was maximum with high biocompatibility, and no signs of systemic toxicity during the entire study period of 35 days (Geng et al., 2017).

Triple-negative breast cancer is one of the most challenging cancer conditions for clinical management. HSA nanoparticles with lapatinib cargo inhibited the adhesion, migration, and invasion ability of high brain-metastatic 4T1 cells *in-vitro*. The nanoparticles also enhanced the survival time of brain metastatic cancer infected mice (Wan et al., 2016). Albendazole entrapped in albumin nanoparticles showed selective toxicity towards ovarian cancer cells compared to healthy ovarian epithelial cells. Nano-albendazole efficiently reduced tumor burden and number of ascites cells in an ovarian cancer xenograft model (Noorani et al., 2015). Dasatinib is a tyrosine kinase inhibitor used for the treatment of myeloid leukemia and acute lymphoblastic leukemia, but its clinical use is limited by serious adverse effects. Cell-specific delivery of the drug is achieved by loading it into albumin nanoparticles, which reduced drug-induced endothelial hyperpermeability without compromising on its anti-leukemia efficacy (Dong et al., 2016). Radionuclide Immune albumin nanosphere design exhibited high drug encapsulation ratio with selective affinity in case of α-fetoprotein positive hepatoma and thereby holds promising potential for its use in radiation gene therapy (Lin et al., 2016). Acyclovir-loaded BSA nanoparticles could easily infuse through human corneal epithelial T cells multilayers without eliciting any toxic response on the cells. Therefore these nanoparticles were definitely better ocular delivery options than simple drug solutions (Suwannoi et al., 2017). Liver cancer cells show overexpression of glycyrrhetinic acid receptor. Doxorubicin-loaded albumin nanoparticles were decorated with glycyrrhetinic acid to effect maximum localization and cytotoxicity in liver cancer cells (Qi et al., 2015). In another study, loading of doxorubicin and other anthracycline derivatives in albumin nanoparticles were experimented through different formulation techniques like covalent bonding or surface adsorption (Kimura et al., 2018). All the nanoparticles with diverse therapeutic load showed similar cellular toxicity. The study provided an insight regarding the formulation development aspects of albumin nanoparticles (Figure 2.3).

Apart from the marketed Abraxane®, several albumin nanoparticles with insoluble anticancer drug cargo is under clinical evaluations like docetaxel and rapamycin (under phase II/III clinical trials), heat-shock

Pharmaceutical Applications of Albumin

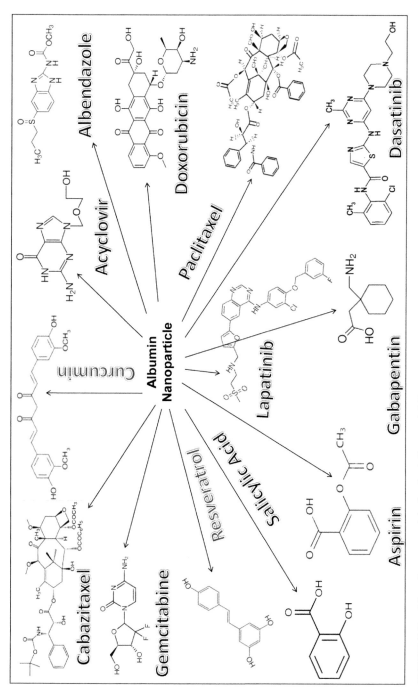

FIGURE 2.3 Albumin nanoparticle for the delivery of different drugs.

protein inhibitor Hsp90 (under phase I/II clinical trials) and tubulin polymerization/topoisomerase I dual inhibitor 5404 (in preclinical stage) (Kouchakzadeh et al., 2015).

Site-specific delivery of drugs for enhancement of drug levels in target sites and reduce unwanted effects is the major goal of all drug delivery approaches. There are several reports where drug carriers made of albumin has been targeted to a particular site to achieve this goal. Covalent conjugation of folate with albumin nanocarriers carrying paclitaxel showed a specific affinity for prostate cancer cells (Zhao et al., 2010). Permeation through the blood-brain barrier is the major hurdle in the brain delivery of drugs. PEGylated albumin nanoparticles conjugated with transferrin showed high localization in brain tissues, and this strategy was successfully explored for the delivery of antiviral drug azidothymidine (Mishra et al., 2006). In another study, glutathione-conjugated BSA nanoparticles were explored for targeted delivery of neuroprotective Asiatic acid. Targeted nanocarriers showed 627.21% drug targeting efficiency compared to Asiatic acid solution establishing its brain targeting potency (Raval et al., 2015). Albumin nanoparticles of two different sizes 10 nm and 200nm loaded with potential anti-angiogenic drug albendazole was evaluated in both *in vitro* and *in vivo* models of ovarian cancer. The smaller nanoparticles exhibited better uptake in SKOV3 ovarian cancer cells, followed by a higher reduction of VEGF levels and tumor burden in mice (Noorani et al., 2015). Another work experimented albumin nanoparticles as a tumor theragnostic agent loaded with chemotherapeutic drug doxorubicin and indocyanine green. Results from *in-vivo* studies revealed greater internalization of the theragnostic nanoparticles in the tumor vicinity than the simple dye solution cementing its claim as a non-invasive tumor monitoring tool.

2.5 ALBUMIN AND ITS CARRIERS: TOXICITY ASPECTS

Albumin is a useful protein and is reported to be nontoxic, non-immunogenic, biocompatible, and biodegradable in nature. However, there are instances where albumin administration in certain clinical conditions has produced adverse effects. For example, Acute Kidney Injury is quite common in patients after cardiac surgery and is a major cause of mortality in such conditions. Results of a retrospective cohort study of 984 patients undergoing cardiac surgery revealed that albumin administration

Pharmaceutical Applications of Albumin 51

influenced a dose-dependent risk for renal injury. The risk induced in patients was nearly two-fold and directly linked to albumin dosage (Frenette et al., 2014). Albumin nanocarriers introduced in mice via oropharyngeal aspiration did not alter TNF-a, and IL-6 levels and neither produced any cellular infiltration. The carriers exhibited a high degree of biocompatibility though slight epithelial damage and neutrophil infiltration was observed at the highest tested dose of 390 µg/mouse (Woods et al., 2015). As discussed earlier HSA has two specific binding sites for different drugs Site I for drugs like warfarin-azapropazone binding site while Site II, for drugs like tryptophan–benzodiazepine. Binding with albumin can even be of irreversible nature and strongly influence both pharmacokinetics and pharmacodynamics of drugs. This phenomenon becomes extremely crucial in the case of combination therapy and for drugs with a narrow therapeutic window (Bertucci et al., 2002). The clinical usefulness of albumin therapy in critically ill patients still possesses some doubtful issues. Reports exist on variable effects of albumin in different patient pools with a critical illness. Insufficient dose administration for anti-inflammatory and antioxidant response and the risk of drug-protein binding in case of combination therapy are other parameters, which contribute towards fluctuations in the result of albumin therapy (Quinlan et al., 2005). Albumin nano/micro-carriers are explored for different disease-specific conditions. Most researchers report that albumin carriers with drug cargo were less cytotoxic compared to the free drug. LD50 value in mice reduced by five holds when free docetaxel was replaced by nanoparticle albumin-bound docetaxel (Desai et al., 2006). However, a fallacy exists for such cytotoxicity determinations as albumin produces misleading results in case of routine evaluations like MTT/XTT assays (Funk et al., 2007).

KEYWORDS

- bovine serum albumin
- drug affinity complex
- epidermal growth factor receptor
- human serum albumin
- nonsteroidal anti-inflammatory drugs

REFERENCES

Ahren, B., (2009). Islet G protein-coupled receptors as potential targets for treatment of type 2 diabetes. *Nat. Rev. Drug Discov.*, *8*, 369–385.

Alba, E., Ciruelos, E., Lopez, R., Lopez-Vega, J. M., Lluch, A., Martín, M., Muñoz, M., Sanchez-Rovira, P., & Seguí, M. Á., (2013). COSTABRAX working group, Liria, M. R., cost-utility analysis of nanoparticle albumin-bound paclitaxel versus paclitaxel in monotherapy in pretreated metastatic breast cancer in Spain. *Expert Review of Pharmacoeconomics & Outcomes Research*, *13*, 381–391.

Andersen, J. T., Dalhus, B., Cameron, J., Daba, M. B., Plumridge, A., Evans, L., Brennan, S. O., Gunnarsen, K. S., Bjørås, M., Sleep, D., & Sandlie, I., (2012). Structure-based mutagenesis reveals the albumin-binding site of the neonatal Fc receptor. *Nat Comm.*, *3*, 610.

Baggio, L. L., Huang, Q., Cao, X., & Drucker, D. J., (2008). An albumin-exendin-4 conjugate engages central and peripheral circuits regulating murine energy and glucose homeostasis. *Gastroenterology*, *134*, 1137–1147.

Bain, V. G., Kaita, K. D., Yoshida, E. M., Swain, M. G., Heathcote, E. J., Neumann, A. U., et al., (2006). A phase 2 study to evaluate the antiviral activity, safety, and pharmacokinetics of recombinant human albumin-interferon alfa fusion protein in genotype one chronic hepatitis C patients. *J. Hepatol.*, *44*, 671–678.

Benet, L. Z., Spahn-Langguth, H., Iwakawa, S., Volland, C., Mizuma, T., Mayer, S., Mutschler, E., & Lin, E. T., (1993). Predictability of the covalent binding of acidic drugs in man. *Life Sci.*, *53*, PL141–146.

Bertucci, C., & Domenici, E., (2002). Reversible and covalent binding of drugs to human serum albumin: Methodological approaches and physiological relevance. *Curr. Med. Chem.*, *9*, 1463–1481.

Birn, H., Fyfe, J. C., Jacobsen, C., Mounier, F., Verroust, P. J., Orskov, H., Willnow, T. E., Moestrup, S. K., & Christensen, E. I., (2000). Cubilin is an albumin binding protein important for renal tubular albumin reabsorption. *J. Clin. Investig.*, *105*, 1353–1361.

Bock, J. B., Bell, A. C., & Herpst, J. Recombinant human albumin-human granulocyte colony stimulating factor for the prevention of neutropenia. U.S. Patent application 2010/0227818 A1, September 9, 2010.

Brekken, R. A., & Sage, E. H., (2000). SPARC, a matricellular protein: At the crossroads of cell-matrix. *Matrix Biol.*, *19*, 569–580.

Bronze-Uhle, E. S., Costa, B. C., Ximenes, V. F., & Lisboa-Filho, P. N., (2017). Synthetic nanoparticles of bovine serum albumin with entrapped salicylic acid. *Nanotechnology, Science and Applications*, *10*, 11–21.

Chaudhury, C., Mehnaz, S., Robinson, J. M., Hayton, W. L., Pearl, D. K., Roopenian, D. C., & Anderson, C. L., (2003). The major histocompatibility complex-related Fc receptor for IgG (FcRn) binds albumin and prolongs its lifespan. *J. Exp. Med.*, *197*, 315–322.

Chawla, S. P., Chua, V. S., Hendifar, A. F., Quon, D. V., Soman, N., Sankhala, K. K., Wieland, D. S., & Levitt, D. J., (2015). A phase 1B/2 study of aldoxorubicin in patients with soft tissue sarcoma. *Cancer*, *121*, 570–579.

Chen, N., Brachmann, C., Liu, X., Pierce, D. W., Dey, J., Kerwin, W. S., Li, Y., Zhou, S., Hou, S., Carleton, M., & Klinghoffer, R. A., (2015). Albumin-bound nanoparticle (nab)

Pharmaceutical Applications of Albumin 53

paclitaxel exhibits enhanced paclitaxel tissue distribution and tumor penetration. *Cancer Chemotherapy and Pharmacology, 76,* 699–712.

Colony Stimulating Factor for the Prevention of Neutropenia, (2010). U. S. Patent application 2010/0227818 A1.

Cui, S., Verroust, P. J., Moestrup, S. K., & Christensen, E. I., (1996). Megalin/gp330 mediates uptake of albumin in renal proximal tubule. *Am. J. Physiol., 271*(4, Pt. 2), 900–907.

Das, S., Bellare, J. R., & Banerjee, R., (2012). Protein-based nanoparticles as platforms for aspirin delivery for ophthalmologic applications. *Colloids and Surfaces B: Biointerfaces., 93,* 161–168.

Demuro, A., Mina, E., Kayed, R., Milton, S. C., Parker, I., & Glabe, C. G., (2005). *J. Biol. Chem., 280,* 17294–17300.

Desai, N., Trieu, V., Yang, A., De, T., Cordia, J., Yim, Z., Ci, S., Louie, L., Grim, B. B., Azoulay, J., Soon-Shiong, P., & Hawkins, M., (2006). Enhanced efficacy and safety of nanoparticle albumin-bound nab-docetaxel versus taxotere. *Proc. Amer. Assoc. Cancer Res., 66,* 1277–1278.

Dong, C., Li, B., Li, Z., Shetty, S., & Fu, J., (2016). Dasatinib-loaded albumin nanoparticles possess diminished endothelial cell barrier disruption and retain potent anti-leukemia cell activity. *Oncotarget., 7,* 49699–49709.

Duttaroy, A., Kanakaraj, P., Osborn, B. L., Schneider, H., Pickeral, O. K., Chen, C., Zhang, G., Kaithamana, S., Singh, M., Schulingkamp, R., Crossan, D., Bock, J., Kaufman, T. E., Reavey, P., Carey-Barber, M., Krishnan, S. R., Garcia, A., Murphy, K., Siskind, J. K., McLean, M. A., Cheng, S., Ruben, S., Birse, C. E., & Blondel, O., (2005). Development of a long-acting insulin analog using albumin fusion technology. *Diabetes, 54,* 251–258.

Elsadek, B. F., & Kratz, F., (2012). Impact of albumin on drug delivery--new applications on the horizon. *J. Control. Release, 157,* 4–28.

Elzoghby, A. O., Samy, W. M., & Elgindy, N. A., (2012). Albumin-based nanoparticles as potential controlled release drug delivery systems. *J. Controlled Release, 157,* 168–182.

Evans, L., Hughes, M., Waters, J., Cameron, J., Dodsworth, N., Tooth, D., Greenfield, A., & Sleep, D., (2010). The production, characterization and enhanced pharmacokinetics of scFv–albumin fusions expressed in Saccharomyces cerevisiae, *Protein Expr. Purif., 73,* 113–124.

Fasano, M., Curry, S., Terreno, E., Galliano, M., Fanali, G., Narciso, P., Notari, S., & Ascenzi, P., (2005). The extraordinary ligand binding properties of human serum albumin. *IUBMB Life, 57,* 787–796.

Foster, J. F., (1977). Some aspects of the structure and the conformational properties of serum albumin. In: Rosenoer, V. M., Oratz, M., & Rothschild, M. A., (eds.), *Albumin Structure, Function and Uses* (p. 53). Oxford: UK.

Frenette, A. J., Bouchard, J., Bernier, P., Charbonneau, A., Nguyen, L. T., Rioux, J. P., Troyanov, S., & Williamson, D. R., (2014). Albumin administration is associated with acute kidney injury in cardiac surgery: A propensity score analysis. *Crit. Care., 18,* 602.

Fu, S., Naing, A., Moulder, S. L., Culotta, K. S., Madoff, D. C., Ng, C. S., Madden, T. L., Falchook, G. S., Hong, D. S., & Kurzrock, R., (2011). Phase I trial of hepatic arterial infusion of nanoparticle albumin-bound paclitaxel: Toxicity, pharmacokinetics, and activity. *Mol. Cancer Ther., 10,* 1300–1307.

Funk, D., Schrenk, H. H., & Frei, E., (2007). Serum albumin leads to false-positive results in the XTT and the MTT assay. *Biotechniques., 43,* 178–182.

Ganong, W. F., (1995). Dynamics of blood and lymph flow. In *Review of Medical Physiology,* 17th ed.; Appleton and Lange: Connecticut, pp. 525.

Geng, T., Zhao, X., Ma, M., Zhu, G., & Yin, L., (2017). Resveratrol-loaded albumin nanoparticles with prolonged blood circulation and improved biocompatibility for highly effective targeted pancreatic tumor therapy. *Nanoscale Research Letters, 12,* 437.

Ghetie, V., Hubbard, J. G., Kim, J. K., Tsen, M. F., Lee, Y., & Ward, E. S., (1996). Abnormally short serum half-lives of IgG in beta 2-microglobulin deficient mice, *Eur. J. Immunol., 26,* 690–696.

Giannoukakis, N., (2003). CJC-1131. *Conju. Chem. Curr. Opin. Investig. Drugs, 4,* 1245–1259.

Gong, G., Pan, Q., Wang, K., Wu, R., Sun, Y., & Lu, Y., (2015). Curcumin-incorporated albumin nanoparticles and its tumor image. *Nanotechnology, 26,* 045603.

Gradishar, W. J., Tjulandin, S., Davidson, N., Shaw, H., Desai, N., Bhar, P., Hawkins, M., & O'Shaughnessy, J., (2005). Phase III trial of nanoparticle albumin-bound paclitaxel compared with polyethylated castor oil-based paclitaxel in women with breast cancer. *J. Clin. Oncol., 23,* 7794–7803.

Guo, L., Peng, Y., Li, Y., Yao, J., Zhang, G., Chen, J., Wang, J., & Sui, L., (2015). Cell death pathway induced by resveratrol-bovine serum albumin nanoparticles in a human ovarian cell line. *Oncology Letters, 9,* 1359–1363.

Halpern, W., Riccobene, T. A., Agostini, H., Baker, K., Stolow, D., Gu, M. L., Hirsch, J., Mahoney, A., Carrell, J., Boyd, E., & Grzegorzewski, K. J., (2002). Albugranin™, a recombinant human granulocyte colony-stimulating factor (G-CSF) genetically fused to recombinant human albumin induces prolonged myelopoietic effects in mice and monkeys. *Pharm. Res., 19,* 1720–1729.

He, X. M., & Carter, D. C., (1992). Atomic structure and chemistry of human serum albumin. *Nature, 358,* 209–215.

Hermansen, K., Davies, M., Derezinski, T., Martinez, R. G., Clauson, P., & Home, P. A., (2006). 26- week, randomized, parallel, treat-to-target trial comparing insulin detemir with NPH insulin as add-on therapy to oral glucose-lowering drugs in insulin-naive people with type 2 diabetes. *Diabetes Care, 29,* 1269–1274.

Hoffman, R. M., & Bouvet, M., (2015). Nanoparticle albumin-bound-paclitaxel: A limited improvement under the current therapeutic paradigm of pancreatic cancer. *Expert Opin. Pharmacother., 16,* 943–947.

Holt, L. J., Basran, A., Jones, K., Chorlton, J., Jespers, L. S., Brewis, N. D., & Tomlinson, I. M., (2008). Anti-serum albumin domain antibodies for extending the half-lives of short-lived drugs. *Protein Eng. Des. Sel., 21,* 283–288.

Home, P., & Kurtzhals, P., (2006). Insulin detemir: From concept to clinical experience. *Expert Opin. Pharmacother., 7,* 325–343.

Hu, Y. J., Liu, Y., Sun, T. Q., Bai, A. M., Lu, J. Q., & Pi, Z. B., (2006). Binding of anti-inflammatory drug cromolyn sodium to bovine serum albumin. *Int. J. Biol. Macromol., 39,* 280–285.

Hurria, A., Blanchard, M. S., Synold, T. W., Mortimer, J., Chung, C. T., Luu, T., Katheria, V., Rotter, A. J., Wong, C., Choi, A., & Feng, T., (2015). Age-related changes in nanoparticle albumin-bound paclitaxel pharmacokinetics and pharmacodynamics: Influence of chronological versus functional age. *The Oncologist, 20,* 37–44.

Jacob, K., Webber, M., Benayahu, D., & Kleinman, H. K., (1999). Osteonectin promotes prostate cancer cell migration and invasion: A possible mechanism for metastasis to bone. *Cancer Res., 59*, 4453–4457.

Junghans, R. P., & Anderson, C. L., (1996). The protection receptor for IgG catabolism is the beta2-microglobulin-containing neonatal intestinal transport receptor. *Proc. Natl. Acad. Sci. U.S.A., 93*, 5512–5516.

Kato, Y., Sakai, N., Baba, M., Kaneko, S., Kondo, K., Kubota, Y., Yao, M., Shuin, T., Saito, S., Koshika, S., Kawase, T., Miyagi, Y., Aoki, I., & Nagashima, Y., (1998). Stimulation of motility of human renal cell carcinoma by SPARC/Osteonectin/BM-40 associated with type IV collagen. *Invasion Metastasis, 18*, 105–114.

Kenanova, V. E., Olafsen, T., Salazar, F. B., Williams, L. E., Knowles, S., & Wu, A. M., (2010). Tuning the serum persistence of human serum albumin domain III: Diabody fusion proteins. *Protein Eng. Des. Sel., 23*, 789–798.

Kim, J. G., Baggio, L. L., Bridon, D. P., Castaigne, J. P., Robitaille, M. F., Jette, L., Benquet, C., & Drucker, D. J., (2003). Development and characterization of a glucagon-like peptide 1-albumin conjugate: The ability to activate the glucagon-like peptide 1 receptor *in vivo*. *Diabetes, 52*, 751–759.

Kimura, K., Yamasaki, K., Nakamura, H., Haratake, M., Taguchi, K., & Otagiri, M., (2018). Preparation and *in vitro* analysis of human serum albumin nanoparticles loaded with anthracycline derivatives. *Chemical and Pharmaceutical Bulletin, 66*, 382–390.

Klein, O., Lynge, J., Endahl, L., Damholt, B., Nosek, L., & Heise, T., (2007). Albumin-bound basal insulin analogs (insulin detemir and NN344): Comparable time-action profiles but less variability than insulin glargine in type 2 diabetes. *Diabetes Obes. Metab., 9*, 290–299.

Kouchakzadeh, H., Safavi, M. S., & Shojaosadati, S. A., (2015). Efficient delivery of therapeutic agents by using targeted albumin nanoparticles. *Adv. Protein Chem. Struct. Biol., 98*, 121–143.

Kragh-Hansen, U., Chuang, V. T., & Otagiri, M., (2002). Practical aspects of the ligand-binding and enzymatic properties of human serum albumin. *Biol. Pharm. Bull., 25*, 695–704.

Kratz, F., Mu, R., Hofmann, I., Drevs, J., & Unger, C., (2000). A novel macromolecular prodrug concept exploiting endogenous serum albumin as a drug carrier for cancer chemotherapy. *J. Med. Chem., 43*, 1253–1256.

Kushwah, V., Agrawal, A. K., Dora, C. P., Mallinson, D., Lamprou, D. A., Gupta, R. C., & Jain, S., (2017). Novel gemcitabine conjugated albumin nanoparticles: A potential strategy to enhance drug efficacy in pancreatic cancer treatment. *Pharmaceutical Research, 34*, 2295–2311.

Lane, T. F., & Sage, E. H., (1994). The biology of SPARC, a protein that modulates cell-matrix interactions. *FASEB J., 8*, 163–173.

Léger, R., Thibaudeau, K., Robitaille, M., Quraishi, O., Van Wyk, P., Bousquet-Gagnon, N., Carette, J., Castaigne, J. P., & Bridon, D. P., (2004). Identification of CJC-1131-albumin bioconjugate as a stable and bioactive GLP-1(7–36) analog. *Bioorg. Med. Chem. Lett., 14*, 4395–4398.

Lin, M., Huang, J., Zhang, D., Jiang, X., Zhang, J., Yu, H., Xiao, Y., Shi, Y., & Guo, T., (2016). Hepatoma-targeted radionuclide immune albumin nanospheres: 131I-antiAFPMcAb-GCV-BSA-NPs. *Analytical Cellular Pathology*, 1–8.

Madsbad, S., Kielgast, U., Asmar, M., Deacon, C. F., Torekov, S. S., & Holst, J. J., (2011). An overview of once-weekly glucagon-like peptide-1 receptor agonists–available efficacy and safety data and perspectives for the future. *Diabetes Obes. Metab.*, *13*, 394–407.

Malik, A. B., (2009). Targeting endothelial cell surface receptors: Novel mechanisms of microvascular endothelial barrier transport. *J. Med. Sci.*, *2*, 13–17.

Melder, R. J., Osborn, B. L., Riccobene, T., Kanakaraj, P., Wei, P., Chen, G., et al., (2005). Pharmacokinetics and *in vitro* and *in vivo* anti-tumor response of an interleukin-2 human serum albumin fusion protein in mice. *Cancer Immunol. Immunother.*, *54*, 535–547.

Merlot, A. M., Kalinowski, D. S., & Richardson, D. R., (2014). Unraveling the mysteries of serum albumin-more than just a serum protein. *Front Physiol.*, *5*, 299.

Merodio, M., Arnedo, A., Renedo, M. J., & Irache, J. M., (2001). Ganciclovir-loaded albumin nanoparticles: Characterization and *in vitro* release properties. *European Journal of Pharmaceutical Sciences*, *12*, 251–259.

Minshall, R. D., Tiruppathi, C., Vogel, S. M., & Malik, A. B., (2002). Vesicle formation and trafficking in endothelial cells and regulation of endothelial barrier function. *Histochem. Cell Biol.*, *117*, 105–112.

Mishra, V., Mahor, S., Rawat, A., Gupta, P. N., Dubey, P., Khatri, K., & Vyas, S. P., (2006). Targeted brain delivery of AZT via transferrin anchored PEGylated albumin nanoparticles. *J. Drug Target*, *14*, 45–53.

Müller, D., Karle, A., Meissburger, B., Höfig, I., Stork, R., & Kontermann, R. E., (2007). Improved pharmacokinetics of recombinant bispecific antibody molecules by fusion to human serum albumin. *J. Biol. Chem.*, *282*, 12650–12660.

Nagaraju, G. P., Dontula, R., El-Rayes, B. F., & Lakka, S. S., (2014). Molecular mechanisms underlying the divergent roles of SPARC in human carcinogenesis. *Carcinogenesis*, *35*, 967–973.

Ngarize, S., Herman, H., Adams, A., & Howell, N., (2004). Comparison of changes in the secondary structure of unheated, heated, and high-pressure-treated beta-lactoglobulin and ovalbumin proteins using Fourier transform Raman spectroscopy and self-deconvolution. *J. Agric. Food Chem.*, *52*, 6470–6477.

Nguyen, A., Reyes, A. E., Zhang, M., McDonald, P., Wong, W. L., Damico, L. A., & Dennis, M. S., (2006). The pharmacokinetics of an albumin-binding Fab (AB.Fab) can be modulated as a function of affinity for albumin. *Protein Eng. Des. Sel.*, *19*, 291–297.

Noorani, L., Stenzel, M., Liang, R., Pourgholami, M. H., & Morris, D. L., (2015). Albumin nanoparticles increase the anticancer efficacy of albendazole in ovarian cancer xenograft model. *Journal of Nanobiotechnology*, *13*, 25.

Oakenfull, D. G., Pearce, R. J., & Burley, R. W., (1997). Protein gelation. In: Damodaran, S., & Paraf, A., (eds.), *Food Proteins and Their Applications* (Vol. 4, p. 111). Marcel Dekker: New York.

Oganesyan, V., Damschroder, M. M., Cook, K. E., Li, Q., Gao, C., Wu, H., & Dall'Acqua, W. F., (2014). Structural insights into neonatal Fc receptor-based recycling mechanisms. *J. Biol. Chem.*, *289*, 7812–7824.

Ottnad, E., Via, D. P., Frubis, J., Sinn, H., Friedrich, E., Ziegler, R., & Dresel, H. A., (1992). Differentiation of binding sites on reconstituted hepatic scavenger receptors using oxidized low-density lipoprotein. *Biochem. J.*, *281*, 745–751.

Peters, T., (1996). *All About Albumin in Biochemistry, Genetics, and Medical Applications*. Academic Press.

Petitpas, I., Bhattacharya, A. A., Twine, S., East, M., & Curry, S., (2001). Crystal structure analysis of warfarin binding to human serum albumin: Anatomy of drug site I. *J. Biol. Chem., 276*, 22804–22809.

Pichler, R. H., Bassuk, J. A., Hugo, C., Reed, M. J., Eng, E., Gordon, K. L., Pippin, J., Alpers, C. E., Couser, W. G., Sage, E. H., & Johnson, R. J., (1996b). SPARC is expressed by mesangial cells in experimental mesangial proliferative nephritis and inhibits platelet-derived-growth-factor-medicated mesangial cell proliferation *in vitro*. *Am. J. Pathol., 148*, 1153–1167.

Pichler, R. H., Hugo, C., Shankland, S. J., Reed, M. J., Bassuk, J. A., Andoh, T. F., et al., (1996a). SPARC is expressed in renal interstitial fibrosis and in renal vascular injury. *Kidney Int., 50*, 1978–1989.

Podhajcer, O. L., Benedetti, L., Girotti, M. R., Prada, F., Salvatierra, E., & Llera, A. S., (2008). The role of the matricellular protein SPARC in the dynamic interaction between the tumor and the host. *Cancer Metastasis Rev., 27*, 523–537.

Qi, W. W., Yu, H. Y., Guo, H., Lou, J., Wang, Z. M., Liu, P., Sapin-Minet, A., Maincent, P., Hong, X. C., Hu, X. M., & Xiao, Y. L., (2015). Doxorubicin-loaded glycyrrhetinic acid modified recombinant human serum albumin nanoparticles for targeting liver tumor chemotherapy. *Molecular Pharmaceutics, 12*, 675–683.

Qu, N., Lee, R. J., Sun, Y., Cai, G., Wang, J., Wang, M., Lu, J., Meng, Q., Teng, L., Wang, D., & Teng, L., (2016). Cabazitaxel-loaded human serum albumin nanoparticles as a therapeutic agent against prostate cancer. *International Journal of Nanomedicine, 11*, 3451–3459.

Quinlan, G. J., Martin, G. S., & Evans, T. W., (2005). Albumin: Biochemical properties and therapeutic potential. *Hepatology, 41*, 1211–1219.

Raval, N., Mistry, T., Acharya, N., & Acharya, S., (2015). Development of glutathione-conjugated Asiatic acid-loaded bovine serum albumin nanoparticles for brain-targeted drug delivery. *J. Pharm. Pharmacol., 67*, 1503–1511.

Rozga, J., Piątek, T., & Małkowski, P., (2013). Human albumin: Old, new, and emerging applications. *Annals of Transplantation, 18*, 205–217.

Ruttala, H. B., & Ko, Y. T., (2015). Liposomal co-delivery of curcumin and albumin/paclitaxel nanoparticle for enhanced synergistic antitumor efficacy. *Colloids and Surfaces B: Biointerfaces, 128*, 419–426.

Sage, H., Vernon, R. B., Funk, S. E., Everitt, E. A., & Angello, J., (1989). SPARC, a secreted protein associated with cellular proliferation, inhibits cell spreading *in vitro* and exhibits Ca + 2-dependent binding to the extracellular matrix. *J. Cell Biol., 109*, 341–356.

Said, N., Frierson, H. F., Sanchez-carbayo, M., Brekken, R. A., & Theodorescu, D., (2013). Loss of SPARC in bladder cancer enhances carcinogenesis and progression. *J. Clin. Invest., 123*, 751–766.

Sand, K. M., Dalhus, B. O., Christianson, G. J., Bern, M., Foss, S., Cameron, J., et al., (2014). Dissection of the FcRn-albumin interface using mutagenesis and anti- FcRn albumin blocking antibodies. *J. Biol. Chem., 289*, 17228–17239.

Schmidt, M. M., Townson, S. A., Andreucci, A. J., King, B. M., Schirmer, E. B., Murillo, A. J., et al., (2013). Crystal structure of an HSA/FcRn complex reveals recycling by competitive mimicry of HSA ligands at a pH-dependent hydrophobic interface. *Structure, 21*, 1966–1978.

Schnitzer, J. E., & Bravo, J., (1993). High-affinity binding, endocytosis, and degradation of conformationally modified albumins. Potential role of gp30 and gp18 as novel scavenger receptors. *J. Biol. Chem.*, *268*, 7562–7570.

Schnitzer, J. E., & Oh, P., (1994). Albondin-mediated capillary permeability to albumin. Differential role of receptors in endothelial transcytosis and endocytosis of native and modified albumins. *J. Biol. Chem.*, *269*, 6072–6082.

Schnitzer, J. E., (1992). gp60 is an albumin-binding glycoprotein expressed by continuous endothelium involved in albumin transcytosis. *Am. J. Physiol.*, *262*(1, Pt. 2), 246–254.

Schnitzer, J. E., (1993). Update on the cellular and molecular basis of capillary permeability. *Trends Cardiovasc Med.*, *3*, 124–130.

Schnitzer, J. E., Sung, A., Horvat, R., & Bravo, J., (1992). Preferential interaction of albumin-binding proteins, gp30 and gp18, with conformationally modified albumins. Presence in many cells and tissues with a possible role in catabolism. *J. Biol. Chem.*, *267*, 24544–24553.

Schnitzer, J., & Oh, P., (1993). Antibodies to the albumin binding protein, albondin, inhibit transvascular transport of albumin in the rat lung. *FASEB Journal*, *7*, A902.

Sudlow, G., Birkett, D. J., & Wade, D. N., (1975). The characterization of two specific drug binding sites on human serum albumin. *Mol. Pharmacol.*, *11*, 824–832.

Suwannoi, P., Chomnawang, M., Sarisuta, N., Reichl, S., & Müller-Goymann, C. C., (2017). Development of acyclovir-loaded albumin nanoparticles and improvement of acyclovir permeation across human corneal epithelial T cells. *Journal of Ocular Pharmacology and Therapeutics*, *33*, 743–752.

Tantra, R., Tompkins, J., & Quincey, P., (2010). Characterization of the de-agglomeration effects of bovine serum albumin on nanoparticles in aqueous suspension. *Colloids Surf. B Biointerfaces.*, *75*, 275–281.

Tijink, B. M., Laeremans, T., Budde, M., Stigter-van Walsum, M., Dreier, T., De Haard, H. J., Leemans, C. R., & Van Dongen, G. A., (2008). Improved tumor targeting of anti-epidermal growth factor receptor nanobodies through albumin binding: Taking advantage of modular nanobody technology. *Mol. Cancer Ther.*, *7*, 2288–2297.

Tiruppathi, C., Finnegan, A., & Malik, A. B., (1996). Isolation and characterization of a cell surface albumin-binding protein from vascular endothelial cells. *Proc. Natl. Acad. Sci. U.S.A.*, *93*, 250–254.

Trüssel, S., Dumelin, C., Frey, K., Villa, A., Buller, F., & Neri, D., (2009). New strategy for the extension of the serum half-life of antibody fragments. *Bioconjugate Chem.*, *20*, 2286–2292.

Vincent, J. L., Russell, J. A., Jacob, M., Martin, G., Guidet, B., Wernerman, J., Roca, R. F., McCluskey, S. A., & Gattinoni, L., (2014). Albumin administration in the acutely ill: What is new and where next? *Critical Care*, *18*, 231.

Volovat, C., Gladkov, O. A., Bondarenko, I. M., Barash, S., Buchner, A., Bias, P., Adar, L., & Avisar, N., (2014). Efficacy and safety of balugrastim compared with pegfilgrastim in patients with breast cancer who are receiving chemotherapy. *J. Clin. Oncol.*, *14*, 101–108.

Wan, X., Zheng, X., Pang, X., Pang, Z., Zhao, J., Zhang, Z., Jiang, T., Xu, W., Zhang, Q., & Jiang, X., (2016). Lapatinib-loaded human serum albumin nanoparticles for the prevention and treatment of triple-negative breast cancer metastasis to the brain. *Oncotarget.*, *7*, 34038–34051.

Wang, W., Ou, Y., & Shi, Y., (2004). AlbuBNP, a recombinant B-type natriuretic peptide and human serum albumin fusion hormone, as a long-term therapy of congestive heart failure. *Pharm. Res.*, *21*, 2105–2111.

Ward, E. S., & Ober, R. J., (2009). Chapter 4: Multitasking by exploitation of intracellular transport functions the many faces of FcRn. *Adv. Immunol., 103*, 77–115.

Werner, T. L., Ray, A., Lamb, J. G., VanBrocklin, M., Hueftle, K., Cohen, A. L., et al., (2017). A phase I study of neoadjuvant chemotherapy with nab-paclitaxel, doxorubicin, and cyclophosphamide in patients with stage II to III breast cancer. *Clinical Breast Cancer, 17*, 503–509.

Williams, A. M., & Dickinson, R. G., (1994). Studies on the reactivity of acyl glucuronides–VI. Modulation of reversible and covalent interaction of diflunisal acyl glucuronide and its isomers with human plasma protein in vitro. *Biochem. Pharmacol., 47*, 457–467.

Wilson, B., Lavanya, Y., Priyadarshini, S. R. B., Ramasamy, M., & Jenita, J. L., (2014). Albumin nanoparticles for the delivery of gabapentin: Preparation, characterization and pharmacodynamic studies. *International Journal of Pharmaceutics, 473*, 73–79.

Woods, A., Patel, A., Spina, D., Riffo-Vasquez, Y., Babin-Morgan, A., De Rosales, R. T., Sunassee, K., Clark, S., Collins, H., Bruce, K., Dailey, L. A., & Forbes, B., (2015). *In vivo* biocompatibility, clearance, and biodistribution of albumin vehicles for pulmonary drug delivery. *J. Control Release, 210*, 1–9.

Zhang, H., Yang, Y., & Steinbrecher, U. P., (1993). Structural requirements for the binding of modified proteins to the scavenger receptor of macrophages. *J. Biol. Chem., 268*, 5535–5542.

Zhao, D., Zhao, X., Zu, Y., Li, J., Zhang, Y., Jiang, R., & Zhang, Z., (2010). Preparation, characterization, and in vitro targeted delivery of folate-decorated paclitaxel-loaded bovine serum albumin nanoparticles. *Int. J. Nanomedicine, 5*, 669–677.

Zheng, Y. R., Suntharalingam, K., Johnstone, T. C., Yoo, H., Lin, W., Brooks, J. G., & Lippard, S. J., (2014). Pt (IV) prodrugs designed to bind non-covalently to human serum albumin for drug delivery. *J. Am. Chem. Soc., 136*, 8790–8798.

CHAPTER 3

Pharmaceutical Applications of Collagen

K. SANGEETHA,[1] A. V. JISHA KUMARI,[2] E. RADHA,[1] and P. N. SUDHA[1]

[1]*Biomaterials Research Lab, Department of Chemistry, D.K.M. College for Women (Autonomous), Vellore, Tamil Nadu, India*

[2]*Department of Chemistry, Tagore Engineering College, Chennai, Tamil Nadu, India*

ABSTRACT

Collagen is the sole most profuse fibrous protein in the animal kingdom, and it was considered as the primary building block of connective tissues and serve as scaffolds for therapeutic agents. The application of collagen was extended both *in vivo* and *in vitro* studies as it was considered as a potential biomaterial with excellent biodegradability, biocompatibility, and weak antigenicity. This chapter emphasized collective information of collagen in terms of its structure, properties, and applications. Also, this chapter addresses the compilation of vast information on the biomedical application of collagen in cancer therapy, drug delivery, tissue engineering, and so on. We have discussed in detail the recent papers covering collagen-based biomaterials which were supposed to become the most prominent material in the field of pharmaceutics in the near future.

3.1 INTRODUCTION

Collagen is principally derived from the word "Kolla" and "genos," which means "glue" and "formation," respectively. Collagen is an insoluble hard, fibrous abounding protein abundantly present in a human, mammalian system and constitutes for 25–35% of the total body weight. It is packed in the body as long, thin fibrils and acts as important fundamental cementing agents in the formation of bones, muscle, skin, and connective tissues (Hopkinson, 1992). Common types of collagen present in a different part of the human body play a vital role in maintaining the suppleness, firmness, and renewability of skin connective tissue. By means of dry weight, collagen comprises the following percentages: sclera with 90%, tendons with 80%, skin with 70–80%, cartilage with 30% and bones and muscle mass with 30%, 1–10%, respectively.

Collagen has a variety of functions like wound healing, revitalization, resurgence of skin tissues, healing, and formation of dentine, bones, cementum, periodontal soft tissue and basement membrane (BM) (Nanci and Bosshardt, 2006). As collagen was weekly immunogenic they can be utilized in drug delivery, ophthalmology, as sponges in curing burns, as mini tablet forms in protein delivery, cosmetic surgery, bone grafting, regeneration of cells in wound, in cardiovascular diseases, in dentines and in eye transplantation. It can be merged with liposome to employ in drug delivery. Collagen nanoparticles are being used more recently in gene delivery (Lee et al., 2001). Enzymatic hydrolysis of collagen yields a bioactive peptide compound called hydrolyzed collagen or collagen hydrolysate with a lower molecular weight of 5000 daltons, and it mainly acts as a messenger in cells for triggering and synthesizing new collagen fibrils and renewing the connective tissues.

The deficiency of collagen may bring about heritable disorders like Ehlers-Dalnos syndrome, Osteogenesis imperfecta, Stickler syndrome, Alport syndrome, Epidermolysis bullosa, Marfan syndrome and autoimmune disorders like Systemic lupus erythematosus, Systemic sclerosis, Oral submucous fibrosis (Rose et al., 2001). The collagen loss was eventually the reason for skin aging (wrinkling and disorganized appearance), as we grow older. Other than this some of the external causes for collagen loss includes severe exposure to the sun, pollution in the air (i.e., continuous exposure to particulate matters in air will leads to loss of collagen), and so on.

3.2 STRUCTURE OF COLLAGEN

Collagen is a multifunctional tissue protein, a major element of the periodontium. They are rod-like structure, which resists stretching and has good tensile strength, which helps to protect against tear or loss of ligaments during stress durations (Perumal et al., 2008). It has a three-way helical structure with 3-α polypeptide chain. Each chain consists of approximately 1000 amino acid residue. The triple helix can hold either homotrimers or

Pharmaceutical Applications of Collagen 63

heterotrimers with continuous or interrupted by certain non-collagenous elements. The diameter and length of the alpha helix was 1.4 nm and 300 nm. The structure of collagen comprises of primary and secondary phases.

In the primary structure of collagen, the basic structure is called as tropocollagen which is a triple helical structure, and each polypeptide chain in the helix has 1056 amino acid residue, and 90% have Gly-A-B amino acid residue (Berisio et al., 2002). Glycine occurs as the third amino acid residue by permitting firm packaging of 3-α chains in the tropocollagen molecule (Bhattacharjee and Bansal, 2005). The basic primary chain structure of collagen was given below:

------- Gly--A-Gly-A-B-Gly-A-B-Gly-A-B-------

where A = proline, and B = hydroxyproline or hydroxylysine.

Collagen has a singular secondary structure, which is helicoidal with higher stretching than the alpha helix. The alpha chains have 3.3 amino acids residues per turn, and they possess left-handed symmetry. The three-polypeptide chains wound around each other to form a triple helical structure with right-handed symmetry (Bella et al., 1994). Interchain hydrogen bonding helps in strengthening of collagen triple helical structure, and the collagen molecules can be further stabilized by intra and intermolecular cross-linking caused by lysine.

3.3 TYPES AND CHARACTERISTIC OF COLLAGEN AS BIOMATERIAL

Around 28 different types of collagens were identified so far, and they were present throughout the body constituting 80–90% of bodybuilding collagen (Shoulder and Raines, 2009). Based on the adoption of structure, collagens were classified as fibrillar and non-fibrillar. Fibrillar collagen includes type I, II, III, V, XI, and non-fibrillar collagen includes (a) Facit – Fibril Associated Collagens with Interrupted Triple Helices) (Type IX, XII, XIV, XIX, XXI) (b) short chain (Type VIII, X) (c) BM (Type IV) (d) Multiplexin (Multiple Triple Helix domains with Interruptions) (Type XV, XVIII) (e) MACIT (Membrane Associated Collagens with Interrupted Triple Helices) (Type XIII, XVII) (f) Other (Type VI, VII). The most common types of collagens and their functions were tabulated in Table 3.1.

The growing quality of collagen as a biomaterial owes to its adaptable multiple applications inherent to collagen's characteristics (Tang et al., 2017).

Some of the important characteristic of collagen includes biodegradability, bioreabsorbability, non-toxic, and support the tissues by acting as extracellular matrix (ECM). They possess extremely high tensile strength, thereby providing greater elasticity and strength to the skin. They can also act as a hemostatic agent and helps in fast recovery of damaged parts by maintaining structural integrity (Buehler, 2006). Scaffolds of collagen can be modified to mimic ECM, which provide support during the regeneration of tissues. The properties of collagen were mainly depending on the raw material from which it was extracted (such as the skin of fish, horse, bovine, etc.) and the method of extraction. The manufacturing processes for effective collagen production include mechanical, physicochemical, and chemical treatments. The different types of collagen and its function were tabulated in Table 3.1.

3.4 EXTRACTION PROCEDURE OF COLLAGEN FROM ANIMALS

3.4.1 FORMATION OF COLLAGEN

Collagen is mainly synthesized in the fibroblast cells of human as procollagen, which serve as a precursor polypeptide (Peterkofsky, 1991). The procollagen, which is synthesized in the ribosomes, is hydroxylated at the amino acid proline and lysine by the enzyme prolyl hydroxylase and lysyl hydroxylase, which needs ascorbic acid (Gorres and Raines, 2010). Additional hydrogen bondings are given by hydroxyl lysyl and hydroxy prolyl residues. In the next step of synthesis, the hydroxyl group in the hydroxyl lysyl amino acid is transferred for glucosyl or galactosyl residues by glucosyl or galactosyltransferase. The alpha chain is created, and then the central part is then altered by combining with other molecules, and the characteristic triple helix is formed, which can then be packed into the vesicles of Golgi apparatus.

3.4.2 FORMATION OF TROPOCOLLAGEN

In Golgi apparatus, the procollagen is modified by post-translational modifications, by addition of polysaccharide which is packed in vesicles and sent to extracellular space. Outside of the Golgi apparatus, the loose ends of the procollagen are removed by collagen peptidase, and the tropocollagen is formed.

TABLE 3.1 Common Types of Collagen and its Functions

Collagen Type	Shape	Location	Origin of Cell	Function
Type I	Three chains with triple helical structure	Skin, bone, tendon, cornea, blood vessels, dentin	Fibroblast, smooth muscle cells, Osteoblast, Odontoblast, chondroblast	Resistance to tension
Type II	Continuous triple helical structure	Hyaline and elastic Cartilage, vitreous humor,	Chondrocytes, Retinal cells	Resistance to intermediate pressure
Type III	Continuous triple helical structure	Lung, Spleen, liver, loose connective tissues, kidney, reticular fibers, blood vessels, Vascular system, papillary layer of dermis,	Fibroblast, smooth muscle cells, reticular cells,	To maintain structure in expansible organs and stability
Type IV	Triple-helical conformations with interruptions of non-helical domains and peptides.	Basement Membrane in various tissues	Endothelial and epithelial cells, muscle cells	To support and to filter
Type V	Continuous triple helical structure	Along with Type I Blood vessel wall, synovium, Corneal stoma, and skeletal muscle	—	—

3.4.3 FORMATION OF COLLAGEN FIBRIL

A copper-containing enzyme lysyl oxidase oxidizes the lysyl and prolyl residues which form an aldehyde group which forms covalent binding with tropocollagen resulting in the organization of collagen fibril (John et al., 1985).

3.4.4 FORMATION OF CROSS-LINKED COLLAGEN

The collagen fibrils formed may undergo certain chemical rearrangements and form covalent cross-linkages to form triple helical collagen. A diagrammatic representation of the mechanism of collagen formation and the cross-linking process was shown in Figures 3.1 and 3.2, respectively.

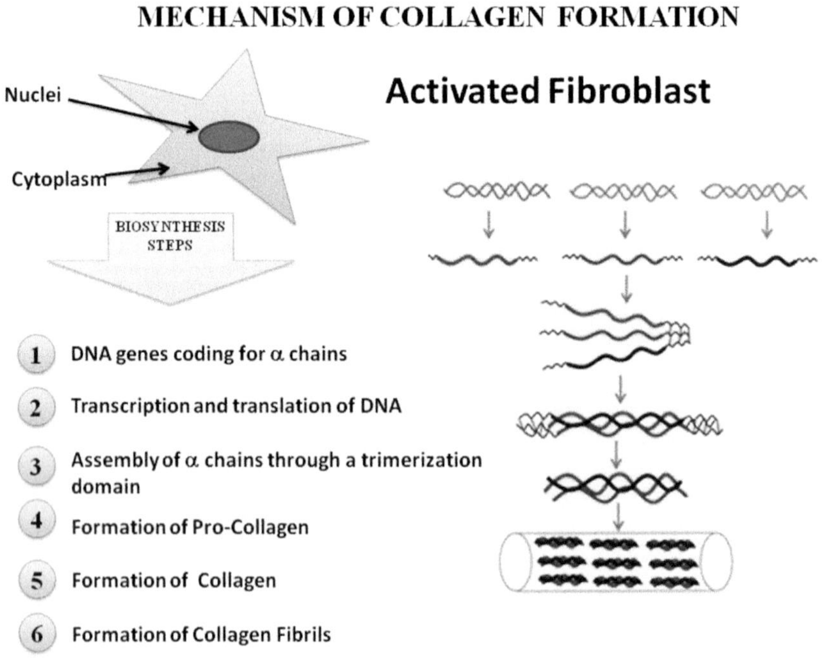

MECHANISM OF COLLAGEN FORMATION

Activated Fibroblast

Nuclei

Cytoplasm

BIOSYNTHESIS STEPS

1 DNA genes coding for α chains

2 Transcription and translation of DNA

3 Assembly of α chains through a trimerization domain

4 Formation of Pro-Collagen

5 Formation of Collagen

6 Formation of Collagen Fibrils

FIGURE 3.1 (See color insert.) Diagrammatic representation of collagen formation (mechanism).
(Adapted from https://www.bulkactives.com/actives/stimulate-collagen-production))

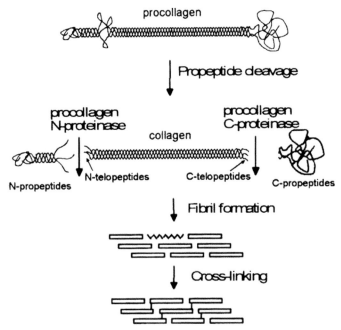

FIGURE 3.2 Schematic representation of cross-linking of collagen. (Reprinted from Pawelec et al., 2016. Copyright (2016) Royal Society of Chemistry. https://creativecommons.org/licenses/by/3.0/)

Montero et al., (1995) and Sato et al., (1986) showed the extraction of collagen from skin, bone fragments, and muscles. Small bits of the above were reacted with 0.8 mol/L of sodium chloride (ratio of collagen and NaCl = 1: 6) for 5 minutes, so that impurities were all removed. The process was repeated for three times with cool distilled water, and again it was mixed with 0.1 mol/L of NaOH (1: 10 ratio) for 3 days which helps in removing the non-collagenous part, and the action of enzyme endogenous protease was inhibited. Collagen extraction was performed using acetic acid by employing three laboratory procedures.

i) **Extraction of Acid Soluble Collagen:** Done by using acetic acid (0.5 mol/L) for 3 days.
ii) **Extraction of Pepsin Soluble Collagen:** Done by addition of pepsin (0.1% and 0.5 mol/L of acetic acid).
iii) **Extraction of Pepsin Soluble Collagen:** Done by centrifuging twice the solution for 20 min at 4°C and the supernatant was salted with 2 mol/L of NaCl for 24 hours at 4°C.

Finally, the precipitated collagen is placed in dialysis membrane, and dialysis is undergone using 0.02 mol/L phosphate buffer. The dialyzed samples were lyophilized and stored in 20°C.

Another interesting extraction procedure of collagen was done by two and half an hour extraction using 0.3 M acetic acid at 90°C. The collagen extracted was washed 2 times using trichloroacetic acid. Another work of collagen isolation was reported by Bentz et al., (1983) from a human from the placenta by the method of Furuto and Miller. Native soluble collagens were prepared by using salt and dilute acid extraction procedures. Addition of inhibitor like EDTA or DFP (diisopropyl fluorophosphate) or phenylmethanesulfonyl fluoride or PMSF (1–5 raM), N-ethylmaleimide, NEM (2–5 mM) which inhibits proteinase enzyme. Extraction is done at 4°C using an appropriate solvent. Now a day's extraction procedure is done under acidic condition has improved the isolation of collagen.

Mokrejs et al., (2009) extracted acid-soluble collagen from the refined material with a volume of 2500 ml of acetic acid at a temperature of 10–35°C using 0.05–0.2 mol of acetic acid. The extraction procedure requires 3–4 steps and the collagen was extracted by centrifuging at 9000 rpm for 5 min at 22°C

3.5 BIOMEDICAL/PHARMACEUTICAL APPLICATIONS OF COLLAGEN

3.5.1 COLLAGEN FOR WOUND DRESSING

Wound healing is a complex process in which the repaired tissues will regenerate through different overlapping phases of (a) blood clotting (hemostasis), (b) inflammation, (c) proliferation and (d) maturation (Martin and Leibovich, 2005). If the healing of wound will occur within 8–12 weeks it was termed as acute wound, and if the mending process was prolonged over a month to years then it was called as chronic wound (Dhivya et al., 2015). Generally, in a chronic wound, the process of healing was so long, and the wound gets easily infected, which will risk the patient morbidity and mortality. Even though a huge number of wound dressing material with commercial name were available in market an ideal chase for wound dressing was still a challenge for researchers and collagen plays a vital role in wound healing as it was the major component in furnishing

Pharmaceutical Applications of Collagen

connective tissues, encouraging the deposition and organization of newly formed collagen, and thereby increase mechanical integrity to strengthen wound will results in foster healing. Collagen-based biomaterials have the ability to stimulate specific cells such as macrophages and fibroblasts, create moisture environment suitable to recover the wound at a faster rate. Here some of the novel combinations of collagen-based wound dressing materials were discussed.

Collagen in different forms such as gel, membrane, sponges were successfully reported as excellent wound dresser. Among the various forms, collagen sponges were more suitable for wound healer as the wet-strength of sponges will allow the suture of soft tissues thereby promoting the growth of new tissues (Boyce et al., 1992). Collagen implants will act as a vehicle to deliver cultured keratinocytes and drugs for replacing affected skin and infected burn wounds. Jinno and his research group (2016) have reported the efficiency of the collagen-gelatin sponge by comparing with the conventional collagen sponge. The prepared collagen implants were placed on the rat wound, and the parameters such as wound area, neoepithelial length, dermis-like tissue area, and the number and area of capillaries were monitored at a time interval of days (one and two weeks). The results revealed that the collagen-gelatin sponges will have sustained release of fibroblast growth factor and thereby accelerating the process of wound healing.

Chu et al., (2018) has constructed collagen-nanomaterial-drug hybrid scaffold comprising of graphene oxide–polyethylene glycol (GO-PEG) followed by incorporation of dietary bioflavonoid, quercetin (3,3',4',5,7-pentahydroxy-flavone, Que) to produce a cellular dermal matrix (GO-PEG-Que) which possess profound antioxidant and anticancer property. The hybrid scaffold showed immense hydrophilicity and delivery of hydrophobic drugs. They also help to promote the stem cell attachment, proliferation, and differentiation, proving the scaffold was effective to repair diabetic wounds.

Tort et al., (2017) have synthesized three-layered wound dressing material using the combination of doxycycline, collagen, chitosan, and alginate as spuned nanofibers, characterized by *in vitro* studies to evaluate the efficacy as wound dresser. This wound dressing material was comprised of three phases (a) first layer was sodium alginate (b) second with chitosan and (c) third layer contains 1% polycaprolactone and 4.5% collagen, the shell comprises 2.5% doxycycline and 2.5% polyethylene oxide. The *in vitro* studies were conducted with keratinocyte cell, and the results of

MTT assay reveal no cytotoxicity with sustained stability at 4°C/ambient humidity. They have reported this wound dressing nanofiber was the best alternative against commercially available products. Edwards and his coworkers (2018) have fabricated collagen matrix by electrochemical deposition that resembles ECM of human skin, and this matrix was cultured in human adipose-derived stem cells and co-culture with keratinocyte. The fabricated collagen matrix possesses high tensile strength and porosity makes them highly suitable for cell growth exhibiting a "cobblestone-like" morphology in stem cell culture and spindle-like morphology in co-culture. This preliminary result supports the effectiveness of matrix to apply as a potential candidate for a variety of cutaneous wounds.

Collagen-based composite wound dressing material was prepared by Xie et al., (2018) in which freeze-drying process was employed to prepare chitosan-collagen-alginate cushion followed by attaching the cushion with polyurethane and its effectiveness was checked by applying the matrix in a rat wound model. The high swelling ability of the composite will provide a moist environment to the wound, thereby promote fast healing of the wound. The *in vitro* results revealed no cytotoxicity and good hemocompatibility which infer this composite as excellent wound dresser. Petersen et al., (2016) have developed novel modifiable collagen–gelatin fleece to examine dermal wounds by applying them in different thickness on minipigs, and the results were reported. The rate of wound recovery was monitored by analyzing closure of affected area per time and histological skin quality, and the best result was reported as collagen-gelatin fleece with a thickness of 150 g/m^2.

Ehterami et al., (2018) have studied the cutaneous wound healing in rat models by using scaffold of poly (ε-caprolactone) (PCL)/collagen impregnated with insulin-chitosan nanoparticles. The sustained release of insulin and cell proliferation was evidenced from MTT assay of L929 cells suggesting the healing capacity was maximum and could be used for clinical application for wound treatment. Ramanathan and his research group (2017) have coated collagen on the scaffold of poly(3-hydroxybutyric acid)-gelatin with the extract of *Coccinia grandis. This* scaffold was tested with male albino Wister rats by creating a wound on the dorsal surface of the rat and has been examined at regular interval of days. The results revealed that the collagen in the scaffold will increase the level of hydroxyproline, hexosamine, uronic acid, thereby promoting the healing efficiency supporting the nature of this material as a wound healer. Another interesting work based on collagen was reported by Karri et al., (2016) by examining *in vitro* drug release and *in vivo* wound healing using the scaffold by loading curcumin and nanochitosan in the collagen-alginate matrix in diabetic rats. The results revealed that the sustained release of curcumin was successfully achieved and the wound closure rate was higher than the control suggesting this scaffold was extended to clinical use for diabetic patients. Shi et al., (2008) had constructed a potent delivery system which delivers angiogenin, blood vessel inducing protein, which helps in the process of angiogenesis. The porous collagen–chitosan scaffold was heparinized and fabricated to deliver growth factor (i.e., ANG) and it was found to be a good dermal substitute, and it is used as an ANG delivery vehicle and can be used in tissue engineering application where enhanced angiogenesis is needed.

Kirubanandan (2017) has developed a bioactive scaffold of porous collagen impregnated with antimicrobial agent ciprofloxacin-loaded gelatin microspheres. This scaffold will deliver the drug in a controlled manner and eradicate wound pathogen and cause immediate tissue regeneration. The release of drug starts within 5 hours, and then a sustained release of drug delivery will follow up to 2 days and heal the full thickness of wounds within a period of 12 days. They have reported this scaffold as a template for the regeneration of epidermal and dermal layers and are used in soft tissue augmentation and wound healing.

3.5.2   AS IMPLANTABLE BIOMATERIALS IN OPTHALMOLOGY

Plastic surgeons used collagen as a primary material for contouring around the eye. Collagen corneal shields were developed as ocular bandages, and there was plenty of research reported by scientists all over the world to monitor the promotion of epithelial cells and stromal healing. These shields were utilized to protect the ocular surface after surgery in traumatic and nontraumatic corneal conditions. Collagen shield was prepared from porcine scleral tissue and was available in the shape of a contact lens in dehydrated form.

Agban and his coworkers (2016) developed a novel nanoparticle impregnated collagen shield to recover from eye disorder "glaucoma." Cross-linked collagen shields were fabricated by adding metal oxide nanoparticles such as titanium dioxide, zinc oxide, and polyvinylpyrrolidone

(PVP) capped zinc oxide. The incorporation of nanoparticle will improve swelling capacity, bioadhesive, and mechanical strength. This shield will result in sustained release of pilocarpine hydrochloride makes them highly efficient compared to normal eye drops used for glaucoma. Zelefsky et al., (2008) prepared a novel biodegradable collagen matrix implant and it was placed in the scleral flap, and the rate of healing was monitored, as this implant was forced to grow in the pores to secrete loose matrixes of connective tissue. The pilot study revealed that the collagen implant was a potential candidate to use in trabeculectomy.

In another similar work reported by Cho and Lee (2015) carried out a clinical study to examine biodegradable Ologen™ collagen matrix to repair scleral thinning. The results revealed the efficiency of the prepared matrix to treat scleral tissue for reconstructing the ocular surfaces. Cillino et al., (2016) have reported a comparative study of collagen matrix implant with mitomycin-C in trabeculectomy followed by postoperation for a period of five-year to evaluate the effectiveness of collagen matrix in trabeculectomy. The outcome of the findings confirmed the efficient of collagen matrix in long term purpose as compared to mitomycin-C.

The first report on corneal collagen cross-linking (CXL) was done by a group of German researchers in the year 2003 to arrest the progression of keratoconus and LASIK-induced ectasia which was generally termed as "post-refractive surgery ectasia. This method will preserve the natural architecture of cornea and thereby prevent the shape of the cornea (as domed shape) without growing steep and irregular. In other words, collagen cross-linking will helps to delay the transplantation of cornea in patients affected by Keratoconus or LASIK. In this technique, riboflavin was used as a photosensitizer, and Ultraviolet-A was employed to maximize intra and interfibrillar covalent bonds formation by photosensitized oxidation (Jankov et al., 2010).

A general method for crosslinking cornea was done by Dresden protocol. This process was carried out by immersing cornea in photosensitizer solution (riboflavin) for 30 minutes and was followed by exposure in UVA light. The drawback of this protocol was it will affect the endothelial cells if the thickness of cornea was less than 400 μm. Hence this protocol will not recommend for healing patients with a thin cornea. To carry out this protocol Hafezi et al., (2009) have swelled thin cornea (examined for 20 patients) to increase the thickness above 400μm prior to the treatment in order to avoid the limitation and these patients were reported with no clinical sign of endothelial damage.

Pharmaceutical Applications of Collagen 73

Hatami-Marbini and Jayaram (2018) have studied an interesting area remains unanswered in corneal crosslinking, i.e., whether the swelling of the cornea before crosslinking treatment will affect the crosslinked corneal biomechanical improvement. They have reported that the process of hydration to increase corneal thickness might not have any significant effect on the extent of biomechanical improvement. Cassagne et al., (2017) has underwent a comparative study of topography-guided corneal collagen cross-linking (TG-CXL) for keratoconus with C-CXL (by Dresden protocol), in a clinical trial of randomly chosen 60 patients in which 30 of them were treated with de-epithelialization focused on the cone, riboflavin application in the cornea followed by irradiation with UV light (TG-CXL) and the remaining 30 patients were treated with Dresden protocol (C-CXL). The improvement in the cornea was regularly monitored for a period of one year. This group has reported that the topography-guided CXL was found to be safer than C-CXL after post operation of one year and it was due to the formation of biological gradient between the cone and the surrounding area which facilitate in the rapid recovery of nerves and cells.

Hersh et al., (2017) have evaluated the CXL to treat keratoconus by examining 205 patients and the post-operation was examined for a period of 1 year suggesting that this method was significant for treating the corneal disorder. Benjamin and Sacha (2017) were also worked on a similar study in which they have examined the keratoconus patient with pigment dispersion syndrome (PDS). They have treated the patients with corneal collagen crosslinking and reported the presence of PDS had no detrimental effect on the outcome of treatment with CXL, and no complication was observed after a regular follow up for 12 months concluding that it was safe to perform CXL in the setting of PDS.

Many researchers have also reported their work by combining collagen crosslinking with additional procedures to enhance its beneficiary effect on eyes in terms of visuality and thereby minimizing the drawbacks. Ohana et al., (2018) have performed a combined study in which collagen crosslinking (CXL) was applied along with photorefractive keratectomy (PRK) for 89 eyes (patients without KC) with different age groups and follow up for one year. This combined treatment will result in improvement in the visual acuity, spherical equivalent, and keratometric values but careful attention with safety measure was needed before approving them as standard protocol.

Kanellopoulos and Asimellis (2015) have performed simultaneous LASIK and CXL to the patients suffering from myopic LASIK with concurrent high-fluence CXL. The treated eyes were monitored for two years, and the report suggest greater stability was achieved in refractive and keratometric outcomes. Rafic and his coworkers (2015) have performed CXL followed by implantation of Visian Toric Implantable Collamer Lens (ICL) for keratoconus and reported greater improvement was achieved in uncorrected distance visual acuity (UDVA) and corrected distance visual acuity (CDVA). Lewis, in his review article (2014), has reported a novel glaucoma implant that was made of ab-interno collagen named as XEN gel stent to regulate aqueous drainage from the eye. This stent is soft, permanent, non-migrating, a subconjunctival implant that shunts fluid from the anterior chamber to the subconjunctival space. The results from the preclinical study and human eye testing support the implant can be used in long term purpose with no irritation in the eye with very low minimal conjunctival tissue disruption.

3.5.3 AS SKIN REPLACEMENT

The presence of collagen was an important form of protein in skin, and as we grow older, the body naturally stops the amount of collagen production which results in wrinkling, facial sagging and rough appearance in the skin. To maintain collagen level stable throughout our life was practically not possible, but we can slow down the effect of aging by taking plenty of nutrients to increase the production of collagen in the body. Collagen supplement will enhance skin elasticity, boost the hydration of the skin, and prevents the damage caused by ultraviolet light.

The largest organ in the human body is skin, and it was frequently damaged by physical and chemical means. Collagen was generally used to regenerate skin and to reestablish skin integrity and functions. The skin tissues comprise of various layers (i.e., dermal, epidermal, etc.) and hence, the artificial skin model should recapitulate three-dimensional architecture of these layers. The basic requirement to be fulfilled by the scaffolds for mimicking skin tissues were evaluated through physicochemical, micro-structural, mechanical, and biological properties.

Song and his coworkers (2017) have developed scaffold from different origins of collagen. A comparative study was made with the fabricated skin chip of collagens and reported that the rat tail model was well fitted in terms of dermal and epidermal layers supporting the formation of *in vitro* skin tissue. Kim et al., (2016) fabricated a novel scaffold of duck's feet collagen/silk fibroin. The duck feet collagen mainly comprises of type I collagen and the porosity, cell penetration and proliferation support the increment in cell viability. The results from animal models reveal that this novel scaffold was highly recommended for tissue regeneration, particularly for skin defects.

Boonrungsiman and his research group (2018) demonstrated the effects of silk fiber based scaffolds by incorporating type I collagen (7.69%) and this scaffold shows excellent stability, water swelling ability with highly porous structure and profound bio-functionality of silk and collagen will increase the adhesion and proliferation of osteoblast and fibroblast cell and reported as good skin substitute. Perumal et al., (2018) synthesized collagen-fucoidan blends for tissue regeneration and *in vitro* studies reveal no cytotoxicity was observed in fibroblast cell with increased proliferation and migration suggesting the blend was an excellent material to apply in tissue regeneration. Chandika et al., (2015) have developed a functional skin tissue using a novel combination of cross-linked fish collagen, sodium alginate, and chitooligosaccharides. This scaffold has an excellent interconnected porous structure with 90% porosity and with the pore size of 160–260 μm. The required pore size and percentage of porosity will help to enhance the cell integrity and cell growth within the 3D structure. They have also undergone *in-vitro* degradation study, which implies the ability of the scaffold to withstand and to support restoration. The *in vitro* degradation results imply mechanical strength in terms of blending materials, cross-linking treatment, and the degree of cross-links suggesting this collagen-based scaffold was a potential candidate for skin replacement.

A similar work using fish collagen was carried out by Cao and his research group (2015) by incorporating freeze-dried scaffold of fish collagen/chitosan/chondroitin sulfate (CS) in poly(lactide-co-glycolide) (PLGA) microsphere. They were evaluated for swelling, degradation, and release profiles suggesting this microsphere will retain its structural integrity and bioactivity and could be used for skin regeneration. Ngan and his coworkers (2015) have carried out an interesting study to regenerate collagen against skin aging by applying nanoemulsions of fullerene on skin twice a day continuously for 28 days. *In vitro* study was conducted using 3T3 fibroblast cell line for 48 hours to evaluate the toxicity and reported no acute toxicity was observed up to the concentration of 1000 µg/mL suggesting the fullerene nanoemulsion was used in large scale production as skin care products. Williamson et al., (2004) used impregnated lyophilized collagen mats with caprolactone as tissue engineering grafts (skin substitutes), and the result showed that these composite films was a suitable substrate for the growth of keratinocyte, fibroblast, and showed immense skin repair.

Ullah et al., (2018) have developed three-dimensional scaffold of chitosan/fish collagen/glycerin and tested with *in vitro* culture of human fibroblasts and keratinocytes. The results revealed that the 3D porous scaffold possesses high porosity and greater cytocompatibility with increased fibroblast proliferation can be effectively used in skin tissue engineering and regeneration. Tangsadthakun et al., (2006) have developed collagen/chitosan scaffold by freeze-drying followed by crosslinking to enhance the stability. These scaffolds were tested with *in-vitro* degradation and cell proliferation using L929 cells. The results revealed the addition of chitosan to collagen will increase its biodegradability, and the scaffold with 30% of chitosan will have enhanced cell proliferation, suggesting them as skin substitute in clinical trials in the future.

3.5.4 AS BONE SUBSTITUTES

Regeneration of bone tissues represents roughly 5,00,000 surgeries for every year in the United States alone. There are numerous clinical reasons to create alternatives in bone tissue engineering, including the requirement for better materials that can be utilized in the reproduction of extensive orthopedic imperfections. A scaffold must have the ability to induce bone formation with the help of surrounding tissues, or it will act as a template for embedded bone cells. Bone implants furnish specialists with an "off the shelf" alternative to help in the mending of imperfections or a spinal combination. The flawless material ought to be embedded into the defect and eventually work as an osteoconductive framework to help and mend the deformity. The number and variety of bone implants that have been produced and clinically trialed worldwide were remarkable.

Salamanca et al., (2018) demonstrated porcine collagen graft material to develop new bones with better critical-size defect space. The excellent physicochemical properties of porcine collagen show the biocompatible nature of the graft and the new bones developed through osteoconductivity was proven to be a reliable graft which may alter GBR treatments in the near future. Hu et al., (2017) systematically assessed the impact of collagen addition with the biological performance of the DCP-rich CPC and its ability to influence osteogenic differentiation. The outcomes from *in vitro* and *in vivo* have proven the enhancement of D1 cell attachment and proliferation, and also elicited the early stage of progenitor cell differentiation from the results of ALP activity and morphological characteristics. A significantly increased bone turnover rate of L4–L5 transverse processes was shown via micro-CT. Overall, they reported that the incorporation of collagen promotes osteoblasts to produce new osteoid matrix, which will become a new treatment strategy in bone imperfection reclamation in spinal infections.

Walsh et al., (2017) revealed the excellence of type I collagen molecule to act as biomaterial and examined bone graft substitute prepared using collagen–tricalcium phosphate and characterize them from a material science perspective, the *in vivo* response was evaluated using critical size cancellous defect model in rabbits, cell, and tissue level utilizing histology and immunohistochemical articulation were associated in the degradation of the components of bone. In their study, the process of bone mending reveal different outcome due to the extraction of collagen from various sources. The method adopted for preparation have direct influence over the resorption profile.

Ho and his coworkers (2016) fabricated a novel biomimetic composite of hydroxyapatite (HA)/β-tricalcium phosphate/collagen with three-dimensional structure and tested as a bone substitute. From their study, they will conclude this collagen composite as an eminent material for bone graft substitute both fundamentally and practically, when utilized in attachment safeguarding after tooth extraction. Another similar work was carried out by Sarikaya and Aydin (2015) in which they have prepared a flexible and biocompatible composite using collagen and β-tricalcium phosphate without using a chemical cross-linker (via dehydrothermal processing, DHT). They have reported that well-interconnected pore structure and homogeneously distributed β-tricalcium phosphate in the network of collagen fibrils was successfully achieved, indicating the ease of handling and biocompatibility of the composite to be used as a bone substitute.

Zanfir and his research group (2016) have synthesized a composite of collagen-HA/barium titanate in which the nanoforms of barium titanate and HA were combined with collagen gels using sol-gel and hydrothermal methods. The *in vitro* study revealed superior osteoinductive properties were observed for all biological samples, which were mainly attributed to the ferromagnetic properties of barium titanate present in the composite. A similar work using collagen and HA was reported by Melnikova et al., (2014) showing outstanding osteogenic properties in *in-vitro* studies suggesting the composite has the tendency to be used as bone graft material.

Campana et al., (2014) also worked on combination of collagen and HA and reported the presence of type I collagen has the ability to enhance osteoblast differentiation along with HA indicating accelerated osteogenesis was achieved when this matrix was embedded with human-like osteoblast cells showing excellent improvement compared to individual polymeric materials in the matrix suggesting them as biomimetic material in bone substitute. Wahl and Czernuszka (2006) have reported that for bone replacement, the most widely chosen materials were collagen and HA as it was the major component in human bone. In their review article they have discussed in detail how collagen and HA will mimic the skeletal bone, along with that they have also reported the importance of factors such as type of collagen, conditions for mineralization, effect of crosslinker, and porosity have direct influence in the results in terms of *in vitro* and *in vivo* studies.

Sela et al., (2000) reported that the incorporation of retinoic acid to collagen at the site of the defected area will upgrade formation of new bone, accomplishing association over the imperfection and prompting its entire repair. Uitterlinden et al., (2001) suggested from his work that the polymorphisms of collagen type I alpha1 and vitamin D receptor as hereditary markers for osteoporotic fracture in women as well as interlocus interaction is an important component of osteoporotic fracture risk.

Chondrocytes present in cartilage will synthesize hyaline and fibrous cartilage by the process called chondrogenesis. Chondrocytes help in the synthesis of ECM called Aggrecan, which forms type II collagen. Damaged cartilage has poor intrinsic regenerative property as they have limited oxygen supply and nutrients, and they also lack blood and lymphatic networks. In combination with mesenchymal stem cell a novel collagen/HA biomimetic scaffold was constructed which induces cell

Pharmaceutical Applications of Collagen 79

differentiation of mesenchymal cells into osteocyte like phenotype cells and help in new bone formation and it showed excellent cytocompatibility (Calabrese et al., 2017).

Zhang et al., (2018) have constructed collagen sponge and had studied the effect of the scaffold in repairing of tendons. The result revealed the collagen sponge will allow the adhesion and proliferation of BMSCs with excellent biocompatibility and good mechano-chemical stimulation suggesting this sponge can be used in tendon replacement surgeries and tendon injury.

3.5.5 AS STENTS AND VASCULAR GRAFT COATINGS

In the current situation, stent implants were considered as the best choice for treating atherosclerosis. However, these implants have few drawbacks of intimal breakage and hyperplasia will result in failures such as in-stent restenosis and late stent thrombosis. Hence in order to minimize these failures, the coating of the stent was generally processed to improve the efficacy as well as to minimize the adverse effect of implants. Collagen was considered as one of the most common materials used as a coating in stents. Many researchers have successfully coated collagen in stent implants and reported a positive feedback with minimal failures.

Zhou and his coworkers (2017) have developed a novel stent for treating tracheomalacia. They were fabricated by processing in digital light through three-dimensional printing followed by collagen I modification. The biocompatibility of stent was evaluated by testing them with cell culture revealing that the stent was hydrophilic, harmless towards cultured bronchial epithelial cells showing more number of live cells confirming the suitability of the stent for tracheomalacia treatment. Kallmes and his coworkers (2001) has published a short communication by constructing stent-grafts made of Polytetrafluoroethylene (PTFE) and type 1 collagen material and have analyzed a comparative study by testing them. These stents were placed in femoral arteries in six mongrel dogs, and the follow up of animals were investigated at regular intervals of time (2 weeks, 6 weeks, and 12 weeks). They have reported that the collagen stent-grafts exhibit a higher rate of patency and suitable as stent grafts.

Chou and his coworker (2017) have reported the risk of stroke was reduced by impregnating collagen on the Dacron patch which initiates thrombosis and promotes restenosis due to the activation of platelets and suggesting the collagen-coated patch was effective to use in carotid endarterectomy. Kilpatrick and Hill (2017) have demonstrated a study on the urethral stent by placing a stent in the proximal urethra of bulldog to resolve urinary obstruction, but the drawback here is after placement of stent will result in severe urinary incontinence. This could be overcome by injecting collagen in penile urethra followed by stent placement will remove the risk of urinary incontinence.

Li and his research group (2014) coated titanium stent with collagen, and hyaluronic acid (HA) in different ratios, and the cell viability was examined with fluorescence staining assay. The coating of collagen and HA prepared with 500 µg/ml and 200 µg/ml displayed significant results in promoting smooth muscle cells (SMCs) to the contractile phenotype. Lin et al., (2010) have coated intravascular stents with anti-CD34 antibody functionalized multilayer of heparin–collagen and the *in vitro* studies were conducted with hemocompatibility tests and cell culture. The outcome of the findings revealed antibody functionalized heparin/collagen will enhance cell attachment and growth of vascular endothelial cells and the *in vivo* results revealed the coated stents possess significant inhibition of neointimal hyperplasia and reduce the rest of in-stent restenosis. Fujiwara et al., (2005) have undergone a comparative study to report the degree of neointimal hyperplasia between endovascular stent grafted with collagen as small-diameter vessels of size lesser than 4 mm and the stent without coating. The results revealed the coated stents have greater hemocompatibility and biocompatibility in the tested rabbits.

Chen et al., (2005) evaluated metallic stents coated with multilayers of collagen and sirolimus followed by crosslinked with genipin and have reported the *in vitro* characteristics, drug release profiles. The role of crosslinker was to provide controlled release of drug and the optimized condition for effective coating was achieved at pH 5 with closely packed adhesion. They have also undergone balloon expansion test to examine the troubles of breaking or peeling of the coated material from the stent. Their findings outlined that the collagen coated stent was efficient without delamination of the coated stent during expansion, and hence, drug-eluting stent was reported as a safe material for sustained sirolimus release. Another interesting coated stent was reported by Nagai et al., (2009), here salmon collagen film was used to cover stent, and the rupture behavior of coating material was tested by expansion method, and it was implanted in carotid arteries of beagles. No rupture was observed after expansion, and this stent was patented with no significant neointimal thickening suggesting them as excellent endovascular stents.

3.5.6 COLLAGEN IN CARDIOLOGY

Heart failure (HF) is a malignant disease affecting around 26 million people worldwide (Yancy et al., 2013). The key mechanism for heart problem is cardiac remodeling, which remains a great challenge for doctors to detect them in the preliminary stage. The main aspects of cardiac remodeling were cardiomyocyte injury and myocardial fibrosis (Heusch et al., 2014). Horn and Trafford (2016) reported that collagen deposition is the main reason for cardiac remodeling, and it will occur most commonly in elderly patients as a result of aging which reduces the heart's diastolic function. Modified collagen plays important roles in several biological processes such as cell differentiation, life/death promotion, and cardiac treatment.

Chemical factors and growth factors have a direct influence in the collagen gene expression and also the cell density, ECM has also been demonstrated to influence collagen synthesis *in vitro*. Yang et al., (1997) reveal exclusive new information that the effect of captopril on collagen biosynthesis appears not to be entirely due to AII formation. However, if it is not AII formation, then the factors that are accountable for modulating collagen biosynthesis, both at the transcription and translation levels have yet to be determined. They recommended that angiotensin II may play an imperative role in cardiac collagen gene expression and may be liable for myocardial fibrosis and thus stiffness.

The alteration in phenotype and cross-linking of collagen appears to have considerable impacts on the ventricular remodeling seen in pressure overload hypertrophy. The collagen changes elucidating diastolic dysfunction and chamber remodeling have been shown in animals. They had to extrapolate these findings to late ventricular remodeling that occurs in hypertensive heart disease in humans. The major inciting stimulus for these changes in collagen type and structure is not clearly known. Limited data is available to understand and explain the transformation of cross-linked to non-cross-linked collagen leading to cardiac remodeling.

Duprez et al., (2018) hypothesized that circulating procollagen type III N-terminal propeptide (PIIINP) and carboxy-terminal telopeptide of type-I collagen (ICTP) predicts incident HF. The narrative of the ICTP assay is parallel to that for PIIINP, using an ICTP-specific antibody. The concentrations in unknown samples are obtained from a calibration curve. Coefficients of variation for internal quality control samples during the main runs were 9.3% for PIIINP high control, 16.5% for PIIINP low control, 6.3% for ICTP high control, and 8.8% for ICTP low control. These may correlate with incident HF and its subtypes.

Horn and Trafford (2016) discussed the potential role of collagen matrix in cardiac remodeling, and recently novel mediators of fibrotic remodeling in age-related cardiac diseases were highlighted. The promising roles for collagen cross-linkers and matrix cellular proteins in post-synthetic collagen modulation, and MMPs as pro-fibrotic mediators give additional complexity to the process of "fibrosis" in aging. These processes will undoubtedly increase, aided by the use of aged animal models in research and imaging technologies such as late gadolinium enhancement CMR imaging, novel mediators of collagen remodeling may become future therapeutic targets in the aged, diseased heart.

The ECM is a multifarious and dynamic entity that drives the formation and improvement of the cardiovascular system, determines critical aspects of cardiovascular performance, which plays key roles in the beginning and succession of abnormal cardiovascular function with aging and disease. A recent pubmed search (accessed August 23, 2016) using the terms ECM and cardiovascular, cardiac, or vascular revealed that this field has been growing annually since the 1990s Driving forces for the increased interest in ECM research include the recognition of biological and pathophysiological importance, improvements in biochemical, cellular, and molecular techniques by which to study this complex unit, and advances in the capacity at microscopic and macroscopic levels to provide greater insight into the dynamically changing ECM. Using combinatorial approaches, ECM research can be explored at greater depths than previously possible. This editorial forms a coalescence of discussions by an ad-hoc panel motivated by a call by the American Heart Association to define important cardiovascular ECM research topic areas. Spinale et al., (2016) reported a cardiac ECM field, which has made remarkable progress, particularly in the past 20 years. The field is poised to answer significant outstanding questions that are fundamental to cardiovascular research. Advancing our understanding of cardiac ECM has the potential to cross-translate to other diseases, because understanding how normal and abnormal cardiac ECM processes occur informs on a broader scale.

Pharmaceutical Applications of Collagen 83

Goudis et al., (2012) reported that degradation and deposition of ECM is a process tightly and enthusiastically regulated by the insubstantial balance between matrix metalloproteinases (MMPs) and tissue inhibitors of matrix metalloproteinases (TIMP 9), which responds to changes in cardiac structure and function during the progression of heart disease. Numerous key issues in which fibrosis alter the atrial function and interacts with other pathophysiological mechanisms to promote the initiation and preservation of atrial fibrillation need to be resolved.

3.5.7 IN CANCER THERAPY

All over the world, cancer is one of the most health threats with an estimated 12.7 million new cases and 7.6 million cancer deaths each year (Jemal et al., 2011). One-third of cancers in high-income societies are caused by various factors relating to food habit, nutrient content, and physical activity. There are specific associations between dietary patterns, foods, body composition or individual nutrients is not simple for the reason that long latent period for cancer development, its complex pathogenesis and the challenge of characterizing the multidimensional aspects of diet and activity over a lifetime. There are many types of cancers which can be treated by vaccination and lifestyle behavior changes. For example, skin cancers could be prevented by avoiding continuous exposure to sun, cervical, and colorectal cancers by screening.

Cancer cells acquire the capacity to migrate to different parts of the body. Tumor cells get the ability to move to different organs, attack different tissues, and colonize these organs, bringing about their spread all through the body. Hence invasion and metastasis of cells can cause death in affected patients. Some genetic and epigenetic mechanisms have been elucidated to overcome the issue of cancer (Rosner et al., 2012; Dawson et al., 2012). Through proteolysis collagen will facilitate the aggressive nature of tumor cells and act as a double-edged weapon in tumor progression, both inhibiting and promoting tumor progression.

We all know that our human body was abundantly comprised of collagen (Brodsky and Persikov, 2005). Later investigations on the collagenous parts of breast cancer biopsy fragments have been centered on an advanced microscopic level to investigate the morphological modification in the deposition of collagen which helps to understand the recognizable proof of the so-called Tumor-Associated Collagen Signatures (TACS), and on the assessment of changes in collagen gene expression. Myllyharju and Kivirikko (2004) reported that the major component in the ECM was collagen and this ECM, in particular, will play a dual role as a suppressor of tumor cells in the early phases yet incomprehensibly as tumor promoters at the later phases of tumor progression.

ECM mainly composed of biomolecules: glucosaminoglycans and fibrous proteins. It plays a dynamic role in plentiful biological processes such as cell differentiation, life/death promotion, and carcinogenesis. Cell-ECM interactions may influence a number of biological activities such as cell proliferation, differentiation, biosynthetic ability, polarity, and loco-motion *via* a number of structurally-different receptors. ECM proteolysis is therefore tightly controlled in normal tissues but typically deregulated in tumors. Extracellular matrix and its constituents are contributing to the knowledge of not only the dynamic relationships between the different components of connective tissues which regulate cell signaling and gene expression in normal conditions, but also the relevant role played by the ECM components in the onset of a considerable number of diseases. Many researchers reported as the Type I collagen is the main structural protein in the interstitial ECM (Kular et al., 2014), Type IV collagen is a key component of the BM, which is found at the basal surface of epithelial and endothelial cells and is essential for tissue polarity (Lu et al., 2012).

This summarized information will help to understand the active interaction between collagen and tumor cells, focusing on changes in physicochemical and biological properties of collagen. A new prototype has been formulated that the intrinsic biomechanical forces in collagen can modulate ECM molecular conformation, producing either protective or destructive molecular and cellular events during tumor progression, depending on the stage of cancer development. Furthermore, the relation-ship between collagen and an immune response is also explored.

3.5.7.1 PROTEOLYSIS OF COLLAGEN

The invasion of cancer was highly facilitated by proteolytic action in which the proteolysis will occur due to the lack of perfect balance between the endogenous proteinases and inhibitors. There were numerous reports in the literature supporting the release of non-collageneous (NC) domains of collagens have the ability to inhibit tumor angiogenesis (Chun et al., 2004).

Pharmaceutical Applications of Collagen

Some of the inhibitor generated by proteolysis include type IV collagen (Kalluri, 2003), endostatin (Arvanitidis and Karsdal, 2016), tumstatin (Hamano et al., 2003) and modulating αvβ3 and αvβ5 integrin signaling (Cooke et al., 2008). These inhibitors influence endothelial cell movement and accordingly, tumor angiogenesis.

3.5.7.2 *INTERACTIONS BETWEEN COLLAGEN AND THE TUMOR-IMMUNE INFILTRATE*

Collagen was traditionally considered as a passive player in tumor progression, but it is currently clear that collagen involves in advancing tumor cells. Collagen changes in tumor microenvironment discharge biomechanical signals, which are detected by both tumor cells and stromal cells, trigger a course of natural occasions. One such evident of collagen role in tumor progression was reported Gouon-Evans et al., (2000) in which mixture of immune cells present in cancer cell were accumulated and migrated within the regions of dense collagen confirming their influences in the immune cell infiltrate. Collagen constitutes the framework of tumor microenvironment and influences tumor microenvironment to such an extent that it directs remodeling of ECM by the degradation of collagen via binding to the glycoprotein Mfge8 (Atabai et al., 2009) and thereby advances the tumor invasion, angiogenesis, attack, and relocation. Ingman et al., (2006) reported the macrophage (an immune cell) will initiate the remodeling and reorganization of the collagen fibers.

3.6 CONCLUSION

Marine sources have started to be explored as reliable and economic sources of collagen. Collagen was considered to be one of the notable biomaterials in pharmaceutical applications as it possesses versatile properties and easily available in abundant. Each type of collagens displays different distinct properties based on their structural features. In this chapter we have discussed in detail the various applications of collagen-based scaffolds including bone substitute, skin regeneration, wound healing, etc., we hope this chapter will cover the recent trends of collagen in the pharmaceutical field.

KEYWORDS

- **bone substitute**
- **cardiology**
- **collagen**
- **opthalmology**
- **sponges**
- **tumor cells**

REFERENCES

Agban, Y., Lian, J., Prabakar, S., Seyfoddin, A., & Rupenthal, I. D., (2016). Nanoparticle cross-linked collagen shields for sustained delivery of pilocarpine hydrochloride. *International Journal of Pharmaceutics, 501*(1/2), 96–101.

Arvantidis, A., & Karsal, M. A., (2016). Chapter 15. Type XV collagen. In book: Biochemistry of collagens, laminins and elastin. *Structure, Function and Biomarkers, 97–99.*

Atabai, K., Jame, S., Azhar, N., Kuo, A., Lam, M., McKleroy, W., Dehart, G., Rahman, S., Xia, D. D., & Melton, A. C., (2009). Mfge8 diminishes the severity of tissue fibrosis in mice by binding and targeting collagen for uptake by macrophages. *J. Clin. Invest., 119*, 3713–3722.

Bella, J., Eaton, M., Brodsky, B., & Berman, H. M., (1994). Crystal and molecular structure of a collagen-like peptide at 1.9A resolution. *Science, 266*(5182), 75–81.

Benjamin, R. L., & Sacha, M., (2017). Corneal collagen crosslinking and pigment dispersion syndrome. *Journal of Cataract & Refractive Surgery, 43*(3), 424–425.

Bentz, H., Morris, N. P., Murray, L. W., Sakai, L. Y., Hollister, D. W., & Burgeson, R. E., (1983). Isolation and partial characterization of a new human collagen with an extended triple-helical structural domain. *Proc. Natl. Acad. Sci., 80*(11), 3168–3172.

Berisio, R., Vitagliano, L., Mazzarella, L., & Zagari, A., (2002). Crystal structure of the collagen triple helix model $[(Pro-Pro-Gly)_{10}]_3$. *Protein Sci., 11*(2), 262–270.

Bhattacharjee. A., & Bansal, M., (2005). Collagen structure: The Madras triple helix and the current scenario. *IUBMB Life, 57*(3), 161–172.

Boonrungsiman, S., Thongtham, N., Suwantong, O., Wutikhun, T., Soykeabkaew, N., & Nimmannit, U., (2018). *An Improvement of Silk-Based Scaffold Properties Using Collagen Type I for Skin Tissue Engineering Applications, 75*(2), 685–700.

Boyce, S. T., Stompro, B. E., & Hansbrough, J. F. (1992). Biotinylation of implantable collagen for drug delivery. *J. Biomed. Mater Res., 26*(4), 547–553.

Brodsky, B., & Persikov, A. V., (2005). Molecular structure of the collagen triple helix. *Adv. Protein Chem., 70,* 301–339.

Buehler, M. J., (2006). Nature designs tough collagen: Explaining the nanostructure of collagen fibrils. *Proc. Natl. Acad. Sci., 103*(33), 12285–12290.

Calabrese, G., Forte, S., Gulino, R., Cefali, F., Figallo, E., Salvatorelli, L., et al., (2017). Combination of collagen-based scaffold and bioactive factors induces adipose-derived mesenchymal stem cells chondrogenic differentiation *in vitro. Front Physiol., 8,* 50.

Campana, V., Milano, G., Pagano, E., Barba, M., Cicione, C., Salonna, G., Lattanzi, W., & Logroscino, G., (2014). Bone substitutes in orthopaedic surgery: From basic science to clinical practice. *J. Mater Sci. Mater Med., 25*(10), 2445–2461.

Cao, H., Chen, M. M., Liu, Y., Liu, Y. Y., Huang, Y. Q., Wang, J. H., Chen, J. D., Zhang, Q. Q., (2015). Fish collagen-based scaffold containing PLGA microspheres for controlled growth factor delivery in skin tissue engineering. *Colloids and Surfaces B: Biointerfaces, 136,* 1098–1106.

Cassagne, M., Pierné, K., Galiacy, S. D., Asfaux-Marfaing, M. P., Fournié, P., & Malecaze, F., (2017). Customized topography-guided corneal collagen cross-linking for keratoconus. *Journal of Refractive Surgery, 33*(5), 290–297.

Chandika, P., Ko, S. C., Oh, G. W., Heo, S. Y., Nguyen, V. T., Jeon, Y. J., et al., (2015). Fish collagen/alginate/chitooligosaccharides integrated scaffold for skin tissue regeneration application. *Int. J. Biol. Macromol., 81,* 504–513.

Chen, M. C., Liang, H. F., Chiu, Y. L., Chang, Y., Wei, H. J., & Sung, H. W., (2005). A novel drug-eluting stent spray-coated with multi-layers of collagen and sirolimus. *Journal of Controlled Release, 108*(1), 178–189.

Cho, C. H., & Lee, S. B., (2015). Biodegradable collagen matrix (Ologen™) implant and conjunctival autograft for scleral necrosis after pterygium excision: Two case reports. *BMC Ophthalmology, 15,* 140.

Chou, D., Tulloch, A., Cossman, D. V., Louis, C. J., Rao, R., Barmparas, G., Mirocha, J., & Wagner, W., (2017). *The Influence of Collagen Impregnation of a Knitted Dacron Patch Used in Carotid Endarterectomy Annals of Vascular Surgery, 39,* 209–215.

Chu, J., Shil, P., Yan, W., Fu, J., Yang, Z., He, C., Deng, X., & Liu, H., (2018). PEGylated graphene oxide-mediated quercetin-modified collagen hybrid scaffold for enhancement of MSCs differentiation potential and diabetic wound healing. *Nanoscale, 10*(20), 9547–9560.

Chun, T. H., Sabeh, F., Ota, I., Murphy, H., McDonagh, K. T., Holmbeck, K., Birkedal-Hansen, H., Allen, E. D., & Weiss, S. J., (2004). MT1-MMP-dependent neovessel formation within the confines of the three-dimensional extracellular matrix. *J. Cell Biol., 167,* 757–767.

Cillino, S., Alessandra, C., Francesco, D. P., Carlo, C., Lucia, L. F., & Giovanni, C., (2016). Biodegradable collagen matrix implant versus mitomycin-C in trabeculectomy: Five-year follow-up. *BMC Ophthalmology,* 16–24.

Cooke, V. G., & Kalluri, R., (2008). Chapter 1: Molecular mechanism of type IV collagen-derived endogenous inhibitors of angiogenesis. *Methods Enzymol., 444,* 1–19.

Dawson, M. A., Kouzarides, T., & Huntly, B. J., (2012). Targeting epigenetic readers in cancer. *N. Engl. J. Med., 367,* 647–657.

Dhivya, S., Padma, V. V., & Santhini, E., (2015). Wound dressings- a review. *Biomedicine, 5,* 22.

Duprez, D. A., Gross, M. D., Kizer, J. R., Ix, J. H., Hundley, W. G., & Jacobs, D. R., (2018). Predictive value of collagen biomarkers for heart failure with and without preserved ejection fraction: MESA (multi-ethnic study of atherosclerosis). *J. Am. Heart Assoc., 7*(5), e007885.

Edwards, N. J., Stone, R., Christy, R., Zhang, C. K., Pollok, B., & Cheng, X., (2018). Differentiation of adipose-derived stem cells to keratinocyte-like cells on an advanced collagen wound matrix. *Tissue and Cell, 53*, 68–75.

Ehterami, A., Salehi, M., Farzamfar, S., Vaez, A., Samadian, H., Sahrapeyma, H., Mirzaii, M., Ghorbani, S., & Goodarzi, A., (2018). *In vitro* and *in vivo* study of PCL/COLL wound dressing loaded with insulin-chitosan nanoparticles on cutaneous wound healing in rats model. *International Journal of Biological Macromolecules, 117*(1), 601–609.

Fessler, J. H., Doege, K. J., Duncan, K. G., & Fessler, L. I., (1985). Biosynthesis of collagen. *Journal of Cellular Biochemistry, 28*(1), 31–37.

Fujiwara, N. H., Kallmes, D. F., Li, S. T., Lin, H. B., & Hagspiel, K. D., (2005). Type 1 collagen as an endovascular stent-graft material for small-diameter vessels: A biocompatibility study. *J. Vasc. Interv. Radiol., 16*(9), 1229–1236.

Gorres, K. L., & Raines, R. T., (2010). Prolyl 4-hydroxylase. *Crit. Rev. Biochem. Mol. Biol., 45*(2), 106–124.

Goudis, C. A., Kallergis, E. M., & Vardas, P. E., (2012). Extracellular matrix alterations in the atria: insights into the mechanisms and perpetuation of atrial fibrillation. *Europace., 14*(5), 623–630.

Gouon-Evans, V., Rothenberg, M. E., & Pollard, J. W., (2000). Postnatal mammary gland development requires macrophages and eosinophils. *Development, 127*, 2269–2282.

Hafezi, F., Mrochen, M., Iseli, H. P., & Seiler, T., (2009). Collagen crosslinking with ultraviolet-A and hypoosmolar riboflavin solution in thin corneas. *J. Cataract Refract. Surg., 35*, 621–624.

Hamano, Y., Zeisberg, M., Sugimoto, H., Lively, J. C., Maeshima, Y., Yang, C., Hynes, R. O., Werb, Z., Sudhakar, A., & Kalluri, R., (2003). Physiological levels of tumstatin, a fragment of collagen IV alpha3 chain, are generated by MMP-9 proteolysis and suppress angiogenesis via alpha V beta3 integrin. *Cancer Cell, 3*, 589–601.

Hatami-Marbini, H., & Jayaram, S. M., (2018). Effect of UVA/riboflavin collagen crosslinking on biomechanics of artificially swollen corneas. *Invest. Ophthalmol. Vis. Sci., 1*(59), 764–770.

Hersh, P. S., DoyleStulting, R., Muller, D., Durrie, D. S., & Rajpal, R. K., (2017). United States multicenter clinical trial of corneal collagen crosslinking for keratoconus treatment. *Ophthalmology, 124*(9), 1259–1270.

Heusch, G., Libby, P., Gersh, B., Yellon, D., Bohm, M., & Lopaschuk, G., (2014). Cardiovascular remodeling in coronary artery disease and heart failure. *Lancet, 383*, 1933–1943.

Ho, K. N., Salamanca, E., Chang, K. C., Shin, T. C., Chang, Y. C., Huang, H. M., et al., (2016). A novel HA/β-TCP – collagen composite enhanced new bone formation for dental extraction socket preservation in beagle dogs. *Materials (Basel), 9*(3), 191.

Hopkinson, I., (1992). Growth factors and extracellular matrix biosynthesis. *J. Wound Care, 1*, 47–50.

Horn, M. A., & Trafford, A. W., (2016). Aging and the cardiac collagen matrix: Novel mediators of fibrotic remodeling. *J. Mol. Cell. Cardiol., 93*, 175–185.

Hu, M. H., Lee, P. Y., Chen, W. C., & Hu, J. J., (2017). Incorporation of collagen in calcium phosphate cements for controlling osseointegration. *Materials (Basel), 10*(8), 910.

Ingman, W. V., Wyckoff, J., Gouon-Evans, V., Condeelis, J., & Pollard, J. W., (2006). Macrophages promote collagen fibrillogenesis around terminal end buds of the developing mammary gland. *Dev. Dyn., 235*, 3222–3229.

Jankov, M. R., Jovanovic, V., Nikolic, L., Lake, J. C., Kymionis, G., & Coskunseven, E., (2010). Corneal collagen cross-linking. *Middle East Afr. J. Ophthalmol., 17*(1), 21–27.

Jemal, A., Bray, F., Center, M. M., Ferlay, J., Ward, E., & Forman, D., (2011). Global cancer statistics. *CA Cancer J. Clin., 61,* 69–90.

Jinno, C., Morimoto, N., Ito, R., Sakamoto, M., Ogino, S., Taira, T., & Suzuki, S., (2016). A comparison of conventional collagen sponge and collagen-gelatin sponge in wound healing. *Bio. Med. Research International,* 1–8.

Kallmes, D. F., Lin, H. B., Fujiwara, N. H., Short, J. G., Hagspiel, K. D., Li, S. T., & Matsumoto, A. H., (2001). Dr. Gary, J. Becker young investigator award: Comparison of small-diameter type 1 collagen stent-grafts and PTFE stent-grafts in a canine model-work in progress. *Journal of Vascular and Interventional Radiology, 12*(10), 1127–1133.

Kalluri, R., (2003). Basement membranes: Structure, assembly and role in tumor angiogenesis. *Nat. Rev. Cancer, 3,* 422–433.

Kanellopoulos, A. J., & Asimellis, G., (2015). Combined laser in situ keratomileusis and prophylactic high-fluence corneal collagen crosslinking for high myopia: Two-year safety and efficacy. *J. Cataract Refract Surg., 41*(7), 1426–1433.

Karri, V. V. S. R., Kuppusamy, G., Venkata, T. S., Mannemala, S. S., Kollipara, R., Devidas, W. A., Mulukutla, S., Satyanarayana, R. K. R., & Malayandi, R., (2016). Curcumin-loaded chitosan nanoparticles impregnated into collagen-alginate scaffolds for diabetic wound healing. *International Journal of Biological Macromolecules, 93,* 1519–1529.

Kilpatrick, S., & Hill, T., (2017). Submucosal collagen injection for management of urinary incontinence following urethral stent placement. *Topics in Companion Animal Medicine, 32*(2), 55–57.

Kim, S. H., Parka, H. S., Lee, O. J., Chao, J. R., Parka, H. J., Lee, J. M., Ju, H. W., Moon, B. M., Parka, Y. R., Song, J. E., Khang, G., & Parka, C. H., (2016). Fabrication of duck's feet collagen–silk hybrid biomaterial for tissue engineering. *International Journal of Biological Macromolecules, 85,* 442–450.

Kirubanandan, S., (2017). Ciprofloxacin-loaded gelatin microspheres impregnated collagen scaffold for augmentation of infected soft tissue. *Asian Journal of Pharmaceutics, 11*(2), 147–158.

Kular, J. K., Basu, S., & Sharma, R. I., (2014). The extracellular matrix: Structure, composition, age-related differences, tools for analysis and applications for tissue engineering. *Journal of Tissue Engineering, 5,* 1–17.

Lee, C. H., Singla, A., & Lee, Y., (2001). Biomedical applications of collagen. *International Journal of Pharmaceutics, 221*(1/2), 1–22.

Lewis, R. A., (2014). Ab internal approach to the subconjunctival space using a collagen glaucoma stent. *J. Cataract Refract Surg., 40*(8), 1301–1306.

Li, J., Zhang, K., Chen, H., Liu, T., Yang, P., Zhao, Y., & Huang, N., (2014). A novel coating of type IV collagen and hyaluronic acid on stent material-titanium for promoting smooth muscle cell contractile phenotype. *Materials Science and Engineering: C., 38*(1), 235–243.

Lin, Q., Ding, X., Qiu, F., Song, X., Fu, G., & Ji, J., (2010). *In situ* endothelialization of intravascular stents coated with an anti-CD34 antibody functionalized heparin–collagen multilayer. *Biomaterials, 31*(14), 4017–4025.

Lu, P., Weaver, V. M., & Werb, X., (2012). The extracellular matrix: A dynamic niche in cancer progression. *J. Cell Biol., 196,* 395–406.

Martin, P., & Leibovich, S. J., (2005). Inflammatory cells during wound repair: The good, the bad and the ugly. *Trends Cell Biol., 15*, 599–607.

Melnikova, S., Zelichenko, E., Zenin, B., Guzeev, V., & Gurova, O., (2014). Bone substitute material on the basis of natural components. *IOP Conference Series: Materials Science and Engineering, 66*(1), 20–25.

Mokrejs, P., Langmaier, F., Mladek, M., Janacova, D., Kolomaznik, K., & Vasek, V., (2009). Extraction of collagen and gelatine from meat industry by-products for food and non-food uses. *Waste Manag. Res., 27*(1), 31–37.

Montero, P., Alvarez, C., Marti, M. A., & Borderias, A. J., (1995). Plaice skin collagen extraction and functional properties. *J. Food Sci., 60*(1), 1–3.

Myllyharju, J., & Kivirikko, K. I., (2004). Collagens, modifying enzymes and their mutations in humans, flies and worms. *Trends Genet, 20*, 33–43.

Nagai, N., Nakayama, Y., Nishi, S., & Munekata, M., (2009). Development of novel covered stents using salmon collagen. *Journal of Artificial Organs, 12*(1), 61–66.

Nanci, A., & Bosshardt, D. D., (2006). Structure of periodontal tissues in health and disease. *Periodontology, 40*, 11–28.

Ngan, C. L., Basri, M., Tripathy, M., Karjiban, R. A., & Abdul-Malek, E., (2015). Skin intervention of fullerene-integrated nanoemulsion in structural and collagen regeneration against skin aging. *European Journal of Pharmaceutical Sciences, 70*(5), 22–28.

Ohana, O., Kaiserman, I., Domniz, Y., Cohen, E., Franco, O., Sela, T., Munzer, G., & Varssano, G., (2018). Outcomes of simultaneous photorefractive keratectomy and collagen crosslinking. *Canadian Journal of Ophthalmology, 53*(5), 523–528.

Pawelec, K. M., Best, M., & Cameron, R. E., (2016). Collagen: A network for regenerative medicine. *J. Mater Chem. B Mater Biol. Med., 4*(40), 6484–6496.

Perumal, R. K., Perumal, S., Thangam, R., Gopinath, A., Ramadass, S. K., Madhan, B., & Sivasubramanian, S., (2018). Collagen-fucoidan blend film with the potential to induce fibroblast proliferation for regenerative applications. *Int. J. Biol. Macromol., 106*, 1032–1040.

Perumal, S., Antipova, O., & Orgel, J. P., (2008). Collagen fibril architecture, domain organization, and triple-helical conformation govern its proteolysis. *Proc. Natl. Acad. Sci., 105*(8), 2824–2829.

Peterkofsky, B., (1991). Ascorbate requirement for hydroxylation and secretion of procollagen: Relationship to inhibition of collagen synthesis in scurvy. *American Journal of Clinical Nutrition, 54*(6), 1135S–1140S.

Petersen, W., Rahmanian-Schwarz, A., Werner, J. O., Schiefer, J., Rothenberger, J., Hübner, G., Schaller, H. E., & Held, M., (2016). The use of collagen-based matrices in the treatment of full-thickness wounds. *Burns., 42*(6), 1257–1264.

Ramanathan, G., Muthukumar, T., & Sivagnanam, U. T., (2017). *In vivo* efficiency of the collagen-coated nanofibrous scaffold and their effect on growth factors and pro-inflammatory cytokines in wound healing. *European Journal of Pharmacology, 814*, 45–55.

Rose, S. P., Ahn, N. U., Levy, H. P., Magid, D., Davis, J., Liberfarb, R. M., Sponseller, P. D., & Francomano, C., (2001). The hip in Stickler syndrome. *J. Pediatr. Orthop., 21*, 657–663.

Rosner, M., & Hengstschlager, M., (2012). Targeting epigenetic readers in cancer. *N. Engl. J. Med., 367*, 1764, 1765.

Salamanca, E., Hsu, C. C., Huang, H. M., Teng, N. C., Lin, C. T., Pan, Y. H., & Chang, W. J., (2018). Bone regeneration using a porcine bone substitute collagen composite *in vitro* and *in vivo*. *Scientific Reports, 8*, Article number 984.

Sarikaya, B., & Aydin, H. M., (2015). Collagen/beta-tricalcium phosphate based synthetic bone grafts via dehydrothermal processing. *Bio. Med. Research International,* 1–9.

Sato, K., Yoshinaka, R., Sato, M., & Ikeda, S., (1986). A simplified method for determining collagen in fish muscle. *Bull. Jpn. Soc. Sci. Fisheries, 52*(5), 889–893.

Sela, J., Kaufman, D., Shoshan, S., & Shani, J., (2000). Retinoic acid enhances the effect of collagen on bone union, following induced non-union defect in guinea pig ulna. *Inflammation Research, 49*(12), 679–683.

Shi, H., Han, C., Mao, Z., Ma, L., & Gao, C., (2008). Enhanced angiogenesis in porous collagen-chitosan scaffolds loaded with angiogenin. *Tissue Eng. Part A., 14*(11), 1775–1785.

Shoulders, M. D., & Raines, R. T., (2009). Collagen structure and stability. *Annu. Rev. Biochem., 78*, 929–958.

Song, H. J., Lim, H. Y., Chun, W., Choi, K. C., Sung, J. H., & Sung, G. Y., (2017). Fabrication of a pumpless, microfluidic skin chip from different collagen sources. *Journal of Industrial and Engineering Chemistry, 56*(25), 375–381.

Spinale, F. G., Frangogiannis, N. G., Hinz, B., Holmes, J. W., Kassiri, Z., & Lindsey, M. L., (2016). Crossing into the next frontier of cardiac extracellular matrix research. *Circulation Research, 119*, 1040–1045.

Tang, S. S., Mohad, V., Gowda, M., & Thibeault, S. L., (2017). Insights into the role of collagen in vocal fold health and disease. *Journal of Voice, 31*(5), 520–527.

Tangsadthakun, C. Kanokpanont, S., Sanchavanakit, N., Banaprasert, T., & Damrongsakkul, S., (2006). Properties of collagen/chitosan scaffolds for skin tissue engineering. *Journal of Metals, Materials and Minerals, 16*(1), 37–44.

Tort, S., Acartürk, F., & Beşikci, A., (2017). Evaluation of three-layered doxycycline-collagen loaded nanofiber wound dressing. *International Journal of Pharmaceutics, 529*(1/2), 642–653.

Uitterlinden, A. G., Burger, H., & Huang, Q., (1998). Relation of alleles of the collagen type ialpha1 gene to bone density and the risk of osteoporotic fractures in postmenopausal women. *N. Engl. J. Med., 338*, 1016–1021.

Ullah, S., Zainol, I., Chowdhury, S. R., & Fauzi, M. B., (2018). Development of various composition multicomponent chitosan/fish collagen/glycerin 3D porous scaffolds: Effect on morphology, mechanical strength, biostability and cytocompatibility. *Int. J. Biol. Macromol., 111*, 158–168.

Wahl, D. A., & Czernuszka, J. T., (2006). Collagen-hydroxyapatite composites for hard tissue repair. *Eur. Cell Mater., 28*(11), 43–56.

Walsh, W. R., Oliver, R. A., Christou, C., Lovric, V., Walsh, E. R., Prado, G. R., & Haider, T., (2017). Critical size bone defect healing using collagen-calcium phosphate bone graft materials. *PLoS One., 12*(1), e0168883.

Williamson, N. M. R., Khammo, N., Adams, E. F., & Coombes, A. G. A., (2004). Composite cell support membranes based on collagen and polycaprolactone for tissue engineering of skin. *Biomaterials, 25*, 4263–4271.

Xie, H., Chen, X., Shen, X., He, Y., Chen, W., Luo, Q., et al., (2018). Preparation of chitosan-collagen-alginate composite dressing and its promoting effects on wound healing. *International Journal of Biological Macromolecules, 107*, 93–104.

Yancy, C. W., Jessup, M., Bozkurt, B., Butler, J., Casey, D. E., & Drazner, M. H., (2013). ACCF/AHA guideline for the management of heart failure: Executive summary: A report of the American College of Cardiology Foundation/American Heart Association Task Force on practice guidelines. *Circulation, 128,* 1810–1852.

Yang, C. M., Kandaswamy, V., Young, D., & Sen, S., (1997). Changes in collagen phenotypes during progression and regression of cardiac hypertrophy. *Cardiovascular Research, 36*(2), 236–245.

Zanfir, A. V., Voicu, G., Busuioc, C., Jinga, S. I., Albu, M. G., & Iordache, F., (2016). New Coll–HA/BT composite materials for hard tissue engineering. *Materials Science and Engineering: C., 62,* 795–805.

Zelefsky, J. R., Hsu, W. C., & Ritch, R., (2008). Biodegradable collagen matrix implant for trabeculectomy. *Expert Review of Ophthalmology, 3*(6), 613–617.

Zhang, B., Luo, Q., Deng, B., Morita, Y., Ju, Y., & Song, G., (2018). Construction of tendon replacement tissue based on collagen sponge and mesenchymal stem cells by coupled mechano-chemical induction and evaluation of its tendon repair abilities. *Acta Biomater., 74,* 247–259.

Zhou, G., Han, Q., Tai, J., Liu, B., Zhang, J., Wang, K., et al., (2017). Digital light procession three-dimensional printing acrylate/collagen composite airway stent for tracheomalacia. *Journal of Bioactive and Compatible Polymers, 32*(4), 1–14.

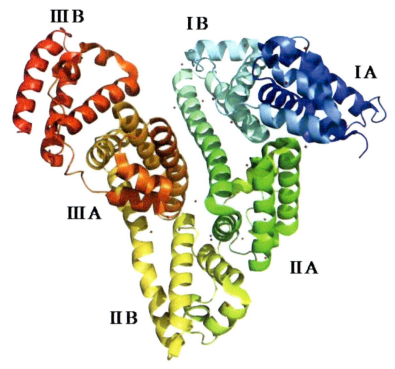

FIGURE 2.1 The crystal structure of human serum albumin in complex with stearic acid. The three domains of albumin are shown in deep blue (IA), light blue (IB), green (IIA), yellow (IIB), light orange (IIIA), and deep orange (IIIB) (PDB 1e7e).

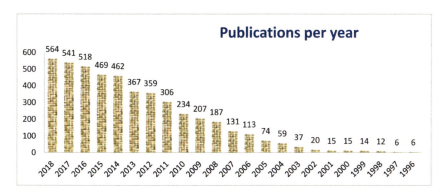

FIGURE 2.2 Year-wise publications as PubMed search on the term "albumin nanoparticles."

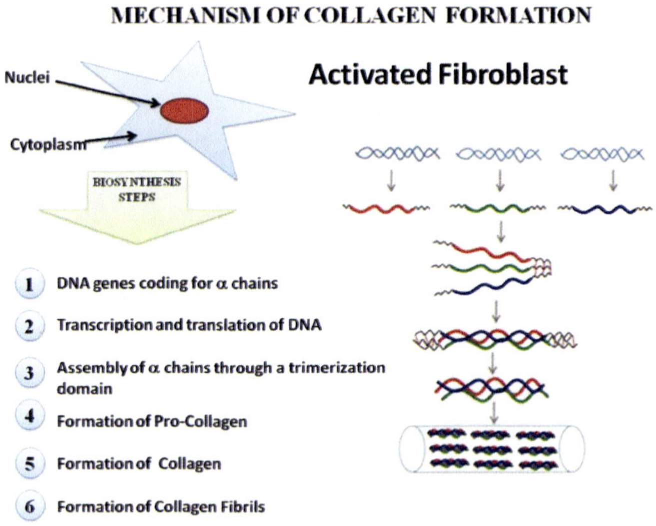

MECHANISM OF COLLAGEN FORMATION

BIOSYNTHESIS STEPS

1. DNA genes coding for α chains
2. Transcription and translation of DNA
3. Assembly of α chains through a trimerization domain
4. Formation of Pro-Collagen
5. Formation of Collagen
6. Formation of Collagen Fibrils

FIGURE 3.1 Diagrammatic representation of collagen formation (mechanism). (Adapted from https://www.pinterest.com/pin/389561436500874075/)

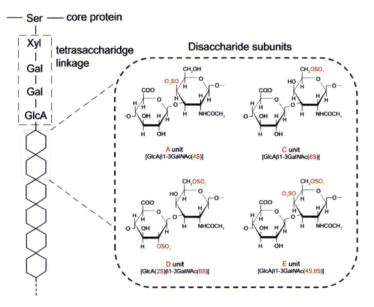

FIGURE 5.1 A schematic diagram showing the structure of CS. CS-glycosaminoglycans are attached to the serine residue on the core protein via a tetrasaccharide linkage.
Source: Kwok et al., (2012); Copyright © 2012 with permission from Elsevier B.V.

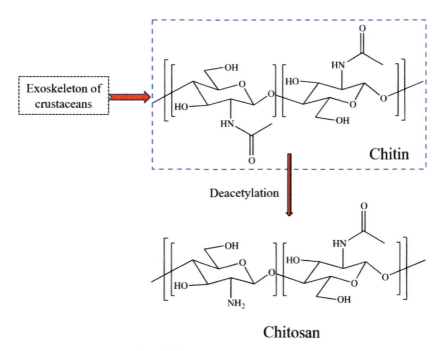

FIGURE 6.2 Deacetylation of chitin to chitosan.

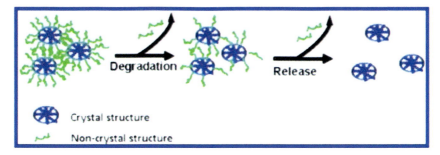

FIGURE 6.5 Degradation mechanism of silk fibroin. Degradation of non-crystal or unstable crystal structures in enzyme solutions which results in the formation of free crystal structure. Then, the crystal structure was dissolved in enzyme solutions.
Reprinted with permission from ACS publication (Lu et al., 2011)

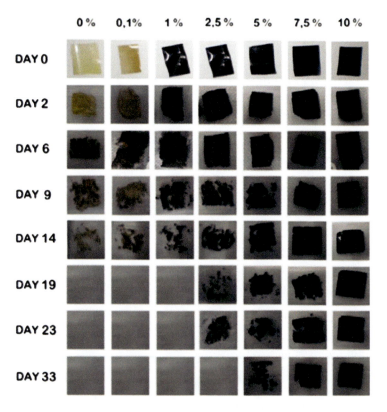

FIGURE 6.6 Macroscopic appearance of the buried soy protein films with 0%; 0.1%; 1%; 2.5%; 5%; 7.5% and 10% (w/w) of genipin after 0, 2, 6, 9, 14, 19, 23 and 33 days. Reprinted with permission from Elsevier (González et al., 2011)

CHAPTER 4

Pharmaceutical Applications of Gelatin

VISHAL GIRDHAR, SHALINI PATIL, SUNIL KUMAR DUBEY, and GAUTAM SINGHVI

Department of Pharmacy, Birla Institute of Technology and Science (BITS), Pilani, Rajasthan–333031, India

ABSTRACT

Natural polymers or biopolymers are always the preferred class of polymers due to their easy availability, eco-friendly nature, biocompatibility, economic as well as their easy modification using certain chemical reagents. Gelatin is a naturally occurring biodegradable and multifunctional biopolymer. Gelatin is a mixture of proteins and peptides obtained by partial hydrolysis of collagen which is extracted from bones and skin trimmings of certain animals. Gelatin is rich in proline, hydroxyproline, and glycine in its chain. Its most common use is in the food industry as a thickening agent and as an emulsifier. Pharmaceutically it finds its application in the hard and soft gelatin capsules. Due to its swelling property, gelatin has found its use in drug delivery by making a hydrogel. Also, it can easily form a film and can be used in the microencapsulation and nanoparticles formation. Thiolated, pegylated, antibody anchored, peptide conjugated and cationized gelatin nanoparticles have been used in various aspects of drug delivery that include anticancer, ocular, and pulmonary delivery, protein, enzyme, and gene delivery. Moreover, polymeric, and lipidic complex nanoparticles with gelatin have also been prepared. Gelatin has also been explored in various tissue engineering scaffolds, films, or wounds healing dressings. The proposed chapter will focus on various strategies for gelatin that have been exploited in both the pharmaceutical as well as bioengineering field.

4.1 INTRODUCTION

Gelatin is a thermoreversible form, obtained by denaturation of collagen, which is an important component of the human body (Vijayaraghavan et al., 2009; Haddar et al., 2012). It is a biocompatible and biodegradable fibrous protein, having wide applications in the field of food technology, pharmaceuticals, biomedical, and cosmetics. Properties of gelatin such as elasticity, long-term stability, and consistency, prove its efficiency for its utilization in food and pharmaceutical industries (Burke et al., 1999; Karim and Bhat, 2009; Jongjareonrak et al., 2010). The thermoreversible nature of gelatin makes it a special polymer for sol-gel transformation. Its conversion from sol-to-gel and gel-to-sol at body temperature and below it, respectively, which make it advantageous over other natural gelling agents (Otoni et al., 2012). It is obtained by extraction from skins and bones of land (Vijayaraghavan et al., 2009), aquatic (Haddar et al., 2012) and poultry animals (Sarbon et al., 2013). Considering a research, it is expected that by 2024 the gelatin market will rise high up to 4 billion US dollars, whereas the market volume is expected to show a growth rate of 5.3% from 2016 to 2024. It is also observed that a considerable portion of gelatin is being utilized by pharmaceutical sector and thus in this chapter discussion will be continued about various different sources & processing of gelatin, and different pharmaceutical carriers developed using the same.

4.2 SOURCES OF GELATIN

Gelatin obtained from collagen through hydrolysis process, is mainly of two types i.e., type A and type B. Type of gelatin depends upon the hydrolysis mechanism, acid hydrolyzed are type A while base hydrolyzed are type B. This collagen for the synthesis of gelatin can be obtained from various sources like bones and skin of animals, fishes, insects, etc. Amongst all the different mammals, gelatin obtained from pig (porcine) and cow (bovine) is the most common source (Tabarestani et al., 2010). Recent statistical data showed that in Europe approximate 80% gelatin is obtained from pig skin, about 15% from cattle hide and the rest 5% is contributed by pig and cattle bone sources and some marine sources, mainly fishes. Gelatin extracted from the animal source (pig and cow) are found to be highly prone to various diseases some may be due to the transmission of

Pharmaceutical Applications of Gelatin 95

prions, others may also include bovine spongiform encephalopathy (BSE) and foot-and-mouth disease (FMD) (Jongjareonrak et al., 2005). Because of the earlier mentioned reasons, nowadays, studies are done to extract and characterize gelatin from poultry and aquatic sources.

During the slaughtering and processing of a poultry, many side products are generated which are considered as waste due to no commercial use of the same. These waste products such as skin, bones, etc. of chickens and ducks are being utilized to produce gelatin these days (Almeida et al., 2013). Gelatin obtained from duck feet has been studied for being an alternative to bovine gelatin and is proved to be a very good alternative to the same (Abedinia et al., 2018). Similarly, chicken skin gelatin has also been studied as a replacement for the commercially available gelatin and is proved to be more useful (Sarbon et al., 2013).

Various aquatic sources of gelatin are also identified these days, which mainly include different fishes as a source of gelatin. Various fishes have been rigorously studied for its gelatin extraction from skin, bones, scales, etc., which include catla, rohu (Madhamuthanalli et al., 2014), bighead carp (Tu et al., 2015), catfish (Jongjareonrak et al., 2010), cuttlefish (Jridi et al., 2013), shark (Kittiphattanabawon et al., 2012), etc. Apart from this, people have even worked on other aquatic sources of gelatin, like octopus, which has shown prominent results (Jridi et al., 2015). But, these sources also suffer from different drawbacks like low melting point (lack of stability), chances of allergic reactions, etc. (Sakaguchi et al., 2000). Thus, nowadays studies have also been started on extracting the gelatin from other land animals like a goat (Mad-Ali et al., 2016), yak (Xu et al., 2017), etc., as well as recombinant gelatin (Olsen et al., 2003) preparation, is also been worked upon.

4.3 CHEMISTRY AND MODIFICATIONS OF GELATIN

The composition of gelatin is based on the amino acid composition of the parent collagen; it includes acidic as well as basic functional groups in it. This structural variability depends on its source and extraction procedure adapted, which finally decides the properties of gelatin (Aguirre et al., 2012). Despite many sources of gelatin, its stability and its inefficiency of utilization into long-term delivery systems (carriers) remains a major issue for wider application. Cross-linking of gelatin has been tried to improve its properties to overcome the existing stability and application issues. Crosslinking of gelatin can be done in various ways like physical crosslinking, chemical crosslinking, radiation crosslinking, etc. (Foox and Zilberman, 2015). Physical crosslinking mainly involves external stimuli such as a change in temperature, change in pH, or some bond interactions, which can be considered under physicochemical interactions (Hoare and Kohane, 2008). Some new methods of cross-linking, including dehydro-thermal treatment, ultraviolet treatment, and plasma treatment, have also been investigated. Unfortunately, these advanced methods also lead to surface changes only and end up with a low degree of crosslinking.

Thus, people started working on the utilization of various chemicals for the crosslinking of gelatin which includes chemicals like glutaraldehyde, 1-ethyl-3-(3-dimethylamino propyl) carbodiimide hydrochloride (EDC), diisocyanates, acyl azides, etc. Presence of an excess number of free functional groups, which are open for reaction with any external compound added, makes chemical crosslinking more favorable option (Ratanavaraporn et al., 2010; Rault et al., 1996). Crosslinking with glutaraldehyde is the most common technique adapted, as glutaraldehyde has a good efficiency of stabilization of final crosslinked product. Also, with the change in concentration of the glutaraldehyde, we can get a varied degree of crosslinking as per the requirement of the final product (Bigi et al., 2001).

Due to the observed cytotoxicity of glutaraldehyde, people have moved towards using naturally occurring agents for crosslinking of gelatin. One of these commonly used agents is genipin, obtained from *Genipa americana* and *Gardenia jasminoides*. This naturally occurring crosslinker is found to have the ability to easily crosslink with freely available amino groups, as in gelatin. Besides less cytotoxicity, it is also found to be biocompatible and thus, finds its application in tissue engineering, food industry, drug delivery, etc. (Cui et al., 2014).

In order to avoid additives and carcinogenic agents, newer techniques like irradiation techniques are used for crosslinking of gelatin. Most commonly, gamma radiations are utilized to initiate polymerization and the correspondingly crosslinking process of gelatin (Jaipan et al., 2017).

4.4 PHARMACEUTICAL APPLICATIONS OF GELATIN

Delivering the drug inside the body with maximum bioavailability, protecting the drug from getting degraded, having a controlled delivery so as to maintain the therapeutic concentration are some of the desired properties of an ideal delivery system. Till date, extensive efforts have been put to obtain such a system. Apart from the conventional applications of gelatin as a component of hard and soft gelatin capsules, it has been explored in various other aspects like as a drug delivery carrier system, as shown in Table 4.1. Being a natural polymer, it has the ability to entrap the drug into its matrix without having any significant harmful effect on the body. By altering the chemistry of gelatin, like altering the isoelectric point of gelatin, modifying the gelatin backbone or addition of different natural and synthetic polymers; different ventures based on the requirement have been tried. Being hydrophilic in nature, easy diffusion of the drug into the body fluid can take place.

4.5 GELATIN-BASED PARTICLES

An extensive study of gelatin to be used as a particulate drug delivery vehicle like nanoparticles and microparticles have been done previously. Both the nanoparticles and the microparticles have their respective advantages and drawbacks over each other. Even liposomes are also developed of gelatin with other lipids keeping in mind its unique design and the ability to incorporate both the hydrophilic as well as hydrophobic drugs. Specific targeting can also be done by surface modification of this particulate system. Gelatin hydrogel beds have been used to encapsulate liposomes of different actives.

4.5.1 NANOPARTICLES

Nanoparticles are defined as submicron sized (1–1000 nm) particles. Nanotechnology has been investigated for several aspects, including pharmacotherapy, food technology, agricultural, and cosmetics. In drug delivery, nanotechnology has been extensively investigated for improvement of existing therapy and targeting drug delivery. Nanocarriers have versatile advantages including increased gastric residence time and permeability, protection of the drug from degradation, enhanced drug absorption by facilitating diffusion through the epithelium, by increased surface area and reactivity (Jun et al., 2011). Due to their multitude advantages, they have been employed for the delivery of any kind of activities including hydrophilic drug, hydrophobic drugs, proteins, vaccines, and imaging and diagnostic agents; to the different areas of the body such as brain, lungs, liver, spleen, kidney, and blood vessels. The surface of the nanoparticles or nanocarrier system could be modified for various applications like in cancer cell targeting, specific organelle level targeting, protection of active, increasing circulation time in blood, and antibody targeting.

In order to improve the loading capacity of the carrier system, a matrix-loaded gelatin nanoparticulate colloidal drug delivery system was prepared (Ofokansi et al., 2010). The gelatin solution was incubated first with the drug solution, and then the formation and crosslinking of nanoparticles were performed at neutral pH in order to retain the native structure of drugs and peptides. A gelatin-based graft copolymer of styrene and hydroxyethyl methacrylate was employed for preparing ibuprofen loaded nanoparticles with an objective of getting a sustained release (Haroun et al., 2014).

Administration of zidovudine, an anti-viral drug, has been a major challenge since it requires multiple dosing because of its poor aqueous solubility, short half-life, and serious side effects. Therefore in order to overcome these problems, the drug was incorporated into nanoparticles based on gelatin modified lipids. A sustained release profile in-vitro was obtained from zidovudine. Also, the blood compatibility studies showed that the modified nanoparticles could be used as a drug delivery vehicle (Joshy et al., 2017).

Nanocomposite film of gelatin containing varying concentrations (0, 3, 5, and 10%) of chitin-nanoparticles was prepared wherein glutaraldehyde was used as a cross-linking agent (Sahraee et al., 2017). Their use as an antifungal agent was proved based on the higher surface hydrophobicity properties and reduced water vapor permeability. Formulation with 5% chitin nanoparticles was more efficacious against *Aspergillus niger* and had better mechanical strength as well.

Tuberculosis still remains one of the most contagious infectious diseases to date because of the resistance was shown by the bacterium to various drugs. One of the reasons behind the development of resistance is the poor compliance since the conventional therapies take as long as 9 months. Therefore, to overcome this challenge, Rajan and Raj (2013) developed a novel self-assembled rifampicin loaded nanoparticle system composed of crab shell chitosan conjugated with PLA using PEG and gelatin by emulsion evaporation technique. From the release profile of the rifampicin, they proposed it as a controlled delivery polymeric carrier.

Gelatin can act as good capping agent for various metallic ions. This makes their role important in the biomedical field. In a study conducted by Divya and coworkers (2018), zinc oxide nanoparticles were coated with gelatin to prove their antibacterial activity. They were more effective against gram-negative *Pseudomonas aeruginosa* as compared to gram-positive *Enterococcus faecalis*. They also exhibited antifungal activity against *Candida albicans* when applied as a film. Using blood vessels of the chick embryo, it was also proved that these nanoparticles had anti-angiogenic activity as well. In a similar case, silver nanoparticles were modified with gelatin to overcome their aggregation problem (Sivera et al., 2014). The nanoparticles obtained herein had high stability towards variable pH conditions.

Gelatin nanoparticles also find their application in the delivery of nuclear matter. Gelatin nanoparticles were loaded with an oligonucleotide of NF-κB decoy and then freeze-dried. Sucrose- and trehalose-containing formulation was more stable under accelerated condition (Zillies et al., 2008).

Delivery of nuclear matter into the human body always poses certain challenges like its stability towards pH, enzymes, and antibodies. Especially when it comes to siRNA delivery, various approaches have been tried including cationic lipid complexes, cationic polymer nanoparticles (Zwiorek et al., 2008), and cationic micelleplexes (Zhao et al., 2012a). Therefore, si-RNA delivery via cholamine surface-modified gelatin nanoparticles for silencing the metastasis linked gene in breast cancer cell line has also been tried which showed nearly 45% release in PBS after 24 hours and up to 96% release in collagenase-enriched PBS (Abozeid et al., 2016). Cellular uptake studies showed high transfection efficiency. Expression of the cancer-causing gene to more than 75% was also observed.

Singh and his co-workers administered systemically, gemcitabine loaded gelatin nanoparticles that had a capability of targeting EGFR in an orthotopic pancreatic cancer model (Singh et al., 2016). Nanoparticles were prepared using desolvation technique and coated with either PEG or EGFR targeting peptide. Use of thiolated type-B gelatin helped in disulfide bond cleavage mechanism of drug release, which is a redox-responsive release. Both in-vivo and in-vitro results were found satisfactory in showing cytotoxicity.

Co-delivery of more than one anticancer drug having synergistic combinations has been exploited for the treatment of different cancers. The problem with these drugs from a delivery point of view is that these drugs have different physicochemical properties. Therefore, their combination delivery becomes a challenge. To keep this problem in consideration, a novel strategy for the co-delivery of camptothecin (a hydrophobic drug) and doxorubicin (hydrophilic drug) was developed where gelatine was used as core to encapsulate the hydrophobic agent and subsequently a calcium phosphate salt of the hydrophilic drug was mineralized over this core to form a shell (Li et al., 2015). This way, interaction of the drugs was prevented. There was a poor release of both the drugs at neutral pH 7.4 indicating better stability during shelf life. Also, as the dissolution pH was changed from 7.4 to 5, the release of the hydrophilic drug increased rapidly along with the calcium of the shell. Camptothecin release increased after 2 hours from the dissolution since the calcium surface degradation was a determining factor for its release. In this way, such type of a delivery system can be used remarkably to provide a controlled, sequential release of multiple drugs at different pH values.

Delivery to the brain has always been a tedious task due to the BBB. Polymeric nanoparticles like gelatin can be used for CNS drug delivery. Extracts of cardamom is a natural anticancer agent without side effects. Cardamom extract-loaded gelatin nanoparticles were prepared by a two-step desolvation technique for the treatment of primary brain tumor glioblastoma (Table 4.1). Nearly 70% of the cardamom extract was released from the nanoparticles in PBS in 72 hours (Nejat et al., 2017).

4.5.2 MICROPARTICLES

Microparticles are a particulate system with a size in the range of 1 μm–1000μm. Microparticles of a drug can be in the form of a matrix or reservoir where the drug is surrounded by a wall of polymers of different thickness and degree of permeability. Change in these two parameters helps us altering the release rates for the drug substance. Microparticles can deliver the drug in a sustained or controlled manner for a longer period of time (Wise, 2000). Microspheres are micrometric matrix systems and essentially spherical in shape. These are polymeric entities in which the drug is physically and uniformly dispersed. Whereas microcapsules are micrometric reservoir systems and microcapsules may be spherical or non-spherical in shape.

TABLE 4.1 Different Applications of Gelatin Nanoparticles in Pharmaceutical Drug Delivery

Delivery	Active	Properties	Observations	References
Ocular	Fluorescein	Double cross-linked Chitosan/gelatin-fluorescein conjugated nanoparticles prepared by reverse emulsion technique.	Increased distribution of the nanoparticles in the retina that persisted for a longer period	Moraru et al., 2014
	Moxifloxacin	Positively charged gelatin nanoparticles by modified desolvation technique	Positively charged nanoparticles with narrow particle size. Burst release for 1 hour, then controlled for further 12 hours in-vitro. Non-irritant to corneal eye surface of rabbits and more effective against S. aureus than MoxiGram®	Mahor et al., 2016
	pDNA	Cationized gelatin nanoparticles	Efficient transfection in vitro (human corneal epithelial cell lines), protection of pDNA in the presence of DNase I for 60 mins compared to 5 mins of naked DNA.	Zorzi et al., 2011a
	pMUC5AC (modified human mucin protein)	Cationized gelatin nanoparticles	A significant increase in MUC5AC mRNA expression in corneal cell lines and conjunctiva of rabbit eye compared to naked plasmid control.	Zorzi et al., 2011b
Pulmonary	Insulin	Water-in-water emulsion technique of making nanoparticles employing glutaraldehyde and poloxamer 188	Nanoparticles obtained had an average size of 250 nm and particles with 1:1 concentration of glutaraldehyde and poloxamer 188 promoted absorption. Rapid hypoglycemic effect.	Zhao et al., 2012b

TABLE 4.1 *(Continued)*

Delivery	Active	Properties	Observations	References
	Cisplatin	Biotinylated-EGF-modified gelatin nanoparticle containing cisplatin	The in-vitro activity showed more potent anticancer activity than free cisplatin. Intratumoral injection to SCID mice showed its strong anti-tumor and less cytotoxic activity. Aerosol delivery to mice with lung cancer showed treatment in lower doses.	Tseng et al., 2009
Nutraceutical delivery	Epigallocatechin gallate (EGCG), tannic acid, curcumin, and theaflavin	Gelatin Nanoparticles coated Layer-by-Layer with polyelectrolytes.	Nanoparticle-encapsulated EGCG retained its biological activity and showed similar potency in inhibiting hepatocyte growth factor (HGF)-induced intracellular signaling in the breast cancer cell line MBA-MD-231 as free EGCG.	Shutava et al., 2009
	Cocoa procyanidins (CPs)	Gelatin-chitosan nanoparticles	CPs-gelatin-chitosan in a mass ratio of 0.75:1:0.5 gave nanoparticles of 344.71 nm size and exhibited the same apoptotic effects at lower concentrations compared with the solution preparation.	Zou et al., 2012
Enzyme immobilization	Glucoamylase	Coacervation method based temperature sensitive cross-linking of gelatin nanoparticles.	Temperature-triggered enzyme immobilization and release. The release took place at a temperature higher than 40°C and no release below this temperature.	Gan et al., 2012
	Diastase Alpha-amylase	Silver nanoparticle (AgNPs) doped gum acacia-gelatin-silica nanohybrid	Nanohybrid retained the bio-catalytic activity and is thermally stable. The shelf life of around 30 days at 40°C and reusable for multiple cycles.	Singh and Ahmed, 2012
	Diastase Alpha-amylase	Carboxymethyl cellulose-gelatin-silica nanohybrid	Immobilization did not alter the working condition. The enzyme could be reused for seven consecutive cycles with nearly 85% of initial activity even at the last cycle.	Singh and Ahmed, 2014

For delivering drugs orally to specifically target the intestinal release, a core-shell gelatin–alginate microparticle system was synthesized using the microfluidic technique (Huang et al., 2014). Alginate stabilizes due to loss of water in acidic pH, thus protecting the material that is embedded and would release the drug-carrying gelatin core (capsules of microspheres) in the alkaline pH.

To overcome drug resistance in tuberculosis, a new microsphere technology was made use of (Manca et al., 2013). Herein the micro-spheres were made of gelatin conjugated covalently with isoniazid and rifampicin was then loaded into the core. This fabrication of the gelatin microsphere with isoniazid proved to be a promising pulmonary drug delivery method.

There has been an intense research on the delivery of propranolol hydrochloride and other antihypertensive agents to eliminate one of the many drawbacks present in conventional therapy. The buccal route has the advantage of giving fast and rapid onset of action, bypassing the first pass metabolism, etc. A tablet dosage form of propranolol hydrochloride based on microparticles of chitosan and gelatin was developed (Abruzzo et al., 2015). The microparticles were prepared by spray drying. The tablets were compressed and subsequently tested for mucoadhesion.

Various studies conducted separately prove that glutaraldehyde has some serious skin and respiratory sensitization issues, eye irritancy, and contact dermatitis. To prevent the harmful effect of this gelatin cross-linker, the use of agarose with gelatin was done for the microencapsulation of gallic acid for the treatment of *Aspergillus niger* fungal infection (Lam et al., 2015). Highest encapsulation was obtained when gelatin and agarose were used in a 1:1 ratio. The release of nearly 80% gallic acid was observed until 96 hours of the study in 7.4 pH buffer. Also, the blank microcapsules did not inhibit the fungal cell growth indicating the harmless nature of agarose in long-term skin sensitization. Similarly, in another study, microspheres made of methacrylated modified gelatin have also been used for the delivery of growth factors (recombinant human BMP-4). Methacrylate derivatives of gelatin have been proved less cytotoxic and provide a broad spectrum of cross-linking densities, thus becoming a possible future for the delivery of various other biomac-romolecules (Nguyen et al., 2015).

4.5.3 LIPOSOMES

Liposomes are small bilayered spherical systems comprising phospholipids which has an aqueous core. These have been successfully used for the delivery of both the hydrophilic as well as the hydrophobic drug. It is one of the most advanced delivery systems because of its many advantages like high biocompatibility with the active molecule and its ability to target the site of action. Because of the use of phospholipids in the liposomes, a hydrophilic core is created and the bilayer formed is hydrophobic in nature which is why liposomes have been used for the entrapment of both the hydrophilic as well as the hydrophobic drugs.

A chitosan/gelatin hydrogel system entrapping phosphatidylcholine liposomes loaded with calcein showed highly controlled and prolonged release and greater liposomal stability (Ciobanu et al., 2014). In a recent study, gelatin nanoformulation of low doses of antiretroviral drug stavudine was incorporated into soya-lecithin based liposomes for preventing the residual viremia effect of HIV-1 virus in-vitro by prolonging the release of the drug (Nayak et al., 2017). In another study, a controlled release gelatin membrane system containing usnic acid loaded liposomes for improving the second-degree skin burns on a porcine model was developed that showed better results than silver sulfadiazine ointment and duoDerme® dressing in terms of collagen deposition and maturation of granulation tissue (Nunes et al., 2016).

Glycerol gelatin based pastilles containing rifampicin loaded liposomes were also prepared to increase the drug's systemic absorption via the oral route (Lankalapalli and Tenneti, 2016). An increased circulatory time of chitosan-gelatin coated phosphatidylcholine and cholesterol liposomes of Irinotecan HCl was observed as compared to the free drug while keeping the drug stable (Shende and Gaud, 2009). The importance of using gelatin as interior support in maintaining the structural integrity, entrapment efficiency, and size during reconstitution of the freeze-dried paclitaxel loaded liposomes is also proved (Guan et al., 2015).

4.6 BIOMEDICAL APPLICATIONS

Apart from the pharmaceutical applications, gelatin has been used for its biomedical applications as well. Carefully engineered scaffolds of gelatin may mimic the extracellular matrix (ECM). They have proved a boon to the human welfare majorly for tissue-engineering and wound healing as they can be used cost-effectively. Hydrogels and bioadhesives of gelatin

Pharmaceutical Applications of Gelatin 105

have gained attention in the last decade because of its swelling property in water. Similarly, a trend in the use of electrospun gelatin fibers for wound healing patches and tissue engineering scaffolds has made its impact.

4.6.1 HYDROGELS

Hydrogels are crosslinked polymeric networks that have a tendency to imbibe a high amount of water content still maintaining its structural alignment. Different polymers of both the natural and synthetic origin have been explored for the preparation of hydrogels. Gelatin is one of the most exploited amongst them all since it is natural in origin, is soluble in water, doesn't has an immunogenic effect and can be easily crosslinked. Hydrogels swell up once they come in contact with water; therefore controlled swelling of the polymer could attribute to its different applications. After swelling, the polymer either dissolves or degrades to create pores. Either the release of drugs or growth factor can occur through these pores or living cells may come and proliferate inside these pores. In a study including micro-engineered methacrylated gelatin, it was found that the immortalized human umbilical vein endothelial cells and 3T3 fibroblasts could easily bound to the matrix and proliferated substantially well when they were both fixed to the methacrylated gelatin substrate as well when they were encapsulated in hydrogels of the same. Thus they offer a promising future to be used for creating complex, cell-responsive microtissues like endothelial lined microvasculature (Nichol et al., 2010). Similar work has been done with methacrylated gelatin hydrogels that are photocrosslinkable/photopolymerizable, to be used as a 3D human vascular network (Chen et al., 2012) and transdermal vascular network for tissue engineering based therapies (Lin et al., 2013).

Pierce et al. developed a gelatin hydrogel that had tunable mechanical properties. It could be obtained by varying the amount of cross-linking agent ethyl lysine diisocyanate. The in vitro cell viability and cell proliferation tests of mesenchymal stem cells (MSC) seeded onto this gelatin-hydrogel showed promising results (Pierce et al., 2012). In another study, a tuneable hydrogel was prepared to overcome the cell adhesion, proliferation, and remodeling inside PEG-based hydrogels. For this purpose, a hydrogel comprising a copolymer of PEG-Gelatin methacrylate with different proportions was prepared that offered the desired results when tested with 3T3 fibroblasts (Hutson et al., 2011). A hydrogel sheet comprising of gelatin honey and chitosan was prepared and evaluated for its antibacterial activity (Wang et al., 2012). Results showed a significantly efficient action against *Staphylococcus aureus* and *Escherichia coli* and also assisted in wound healing.

4.6.2 BIOADHESIVES

Due to its multidimensional advantageous characteristics, gelatin has been successfully used as an alternative to staples and sutures for the bio-adhesion either alone or along with other additives like alginate. Gelatin is biocompatible, biodegradable, and non-immunogenic, has polar amino and carboxyl groups for crosslinking and has natural stickiness, which makes it suitable to be used as a bioadhesive especially in pain management, wound healing and antibiotic drug delivery. To increase the crosslinking of gelatin chemically and further the mechanical properties, alginate has been used (Cohen et al., 2013). Carbodiimide was used as a crosslinking agent of gelatin and alginate for the delivery of two pain relieving agents, bupivacaine, and ibuprofen. Bupivacaine the local anesthetic helped in improving the bioadhesive strength of the gelatin-alginate bioadhesive. A burst release of around 44–74% after 6 hours and then a sustained release of up to 99% were observed at 3 days. This implies that it could be used as a good candidate for burns related pain since with the healing of the burn, pain associated with it decreases (Cohen et al., 2014a). The same group of scientists then worked upon the hard tissue adhesives for the bone fixation where they found had calcium phosphate increased the bonding strength of the gelatin-alginate bioadhesive when hydroxyapatite (HA) and tricalcium phosphate were used as ceramics for bone filler (Cohen et al., 2014b). Also, the effect of hemostatic agents, namely tranexamic acid and kaolin on the bioadhesive and bonding strength of the same bioadhesive was also studied (Pinkas and Zilbernan, 2014).

Feng and his co-workers developed an injectable gelatin hydrogel by creating a supramolecular gelatin-β cyclodextrin complex which could help in the successful delivery of encapsulated cells for the in-situ tissue regeneration (Feng et al., 2016). Crosslinking occurred due to the weak host-guest interactions between gelatin and cyclodextrin.

4.6.3   FIBERS

Polymeric fibers have been long used in the biomedical applications for wound healing patches and tissue-engineered matrices (Maleknia et al., 2014). Since they can form a mesh-like, a network that is similar in structure to the natural ECM of the body. Formation of the gelatin fibers can be done in many ways. Melt spinning, wet spinning, gel spinning or dry spinning are some of the techniques by which fibers with bigger diameters are prepared. To produce fibers in the nano range, electrospinning, centrifugal spinning, and solution/melt-blow spinning are the method of choice. The high quality of work has been done on the biomedical applications of gelatin that include fibers. Amongst all these techniques electrospinning method has gained a lot of attention because it helps in producing a porous network that has a high surface-to-volume ratio, which is one of the desired characteristics to increase the solubility of the drug and simultaneously its release profile. Also, the degree of porosity and pore size can be controlled by altering process parameters like temperature, humidity, and solution type.

Gelatin scaffolds for bone tissue regeneration have an advantage over other scaffolds of synthetic origin in not being cytotoxic and providing better mechanical properties when used in conjugation with such synthetic polymer. Kim et al., developed a PLLA/gelatin scaffold that helped in rapid cell population and growth (Kim et al., 2008). Nanofibrous gelatin/apatite composite scaffolds were prepared. Its mechanical and biocompatible studies showed that it could be used as a substitute for the bone ECM. Apatite was found to have a positive osteogenic differentiation effect (Liu et al., 2009). A three-dimensional calcium phosphate nanoparticles modified gelatin/poly (ε-caprolactone) (PCL) fibrous bilayered scaffold was also found to have better bone adhesion properties than gelatin scaffold alone. Calcium phosphate nanoparticles also increased the bone mineralization (Rajzer et al., 2014).

Gelatin and PCL (70:30) hybrid electrospun fibrous membranes have also used a substitute to cartilage. For the purpose of controlling the shape, titanium alloy mold was used, and the membranes were layered over it that consisted of chondrocytes seeded into the membrane matrix (Xue et al., 2013).

Another application of such fibers of gelatin is in wound dressings. To prevent the bacterial infection in severe burns, gentamycin sulfate and ciprofloxacin were loaded in a pH-dependent gelatin/ alginate dialdehyde electrospun fiber that exhibits multiple releases. A complete release of gentamycin sulfate occurred for 6 days, and that of ciprofloxacin went for over 3 weeks inhibiting *Pseudomonas aeruginosa* and *Staphylococcus epidermidis* (Chen et al., 2016).

Various other fields where gelatin related electrospun fibers have been used are as artificial neural implants, (Baiguera et al., 2014) retinal epithelium basement (Gu et al., 2011), corneal ECM (Tonsomboon and Oyen, 2013) and drug delivery into vascular tissues (Wang et al., 2013).

4.7 FUTURE PERSPECTIVES

With the increase in vegetarian population and stringent regulatory directives, it becomes the necessity of time to produce gelatin from non-animal sources. The development of newer technologies like recombinant gelatin production can be an absolute game changer in this regard. The major drawback with gelatin comes with the crosslinking. There is a need in the future to find a non-toxic and inert crosslinking agent that can serve the purpose to the best. Natural origin crosslinkers have a good potential in particulate drug delivery. Exploring different delivery routes using gelatin is open for research. Specific targeting using modified gelatin, especially in tumor tissue and brain, can be investigated. Effective release and distribution of the drug from gelatin carrier based on microenvironment can help overcome various hitches of the already present delivery systems. Three-dimensional cell culture can be used using modified gelatin so that knowledge about disease progression can be obtained.

4.8 CONCLUSION

Upon a brief review of the literature, it can be said that gelatin is a very valuable and functional protein both for the food industry and more importantly, the pharmaceutical industry. Its use is now not limited to cakes, confectionary, deserts, capsules, tablets, and delivery vehicle, but it has made its place in the biomedical field as well. Its application in the tissue engineering and wound healing scaffolds is tremendous. Rather, modifying the gelatin has provided colorful results as per the need. Use of gelatin hydrogels, fibers, and bioadhesive in arthritis, osteoporosis, bone reformation, cell culture studies, retinal, and corneal implants for drug delivery have gained a lot of attention.

KEYWORDS

- **biodegradable**
- **biopolymers**
- **drug delivery**
- **gelatin**
- **natural polymers**

REFERENCES

Abedinia, A., Ariffin, F., Huda, N., & Nafchi, A. M., (2018). Preparation and characterization of a novel biocomposite based on duck feet gelatin as alternative to bovine gelatin. *International Journal of Biological Macromolecules, 109,* 855–862.

Abozeid, S. M., Hathout, R. M., & Abou-Aisha, K., (2016). Silencing of the metastasis-linked gene, AEG-1, using siRNA-loaded cholamine surface-modified gelatin nanoparticles in the breast carcinoma cell line MCF-7. *Colloids and Surfaces B: Biointerfaces, 145,* 607–616.

Abruzzo, A., Cerchiara, T., Bigucci, F., Gallucci, M. C., & Luppi, B., (2015). Mucoadhesive buccal tablets based on chitosan/gelatin microparticles for delivery of propranolol hydrochloride. *Journal of Pharmaceutical Sciences, 104*(12), 4365–4372.

Aguirre-Álvarez, G., Foster, T., & Hill, S. E., (2012). Impact of the origin of gelatins on their intrinsic properties. *CyTA-Journal of Food, 10*(4), 306–312.

Baiguera, S., Del Gaudio, C., Lucatelli, E., Kuevda, E., Boieri, M., Mazzanti, B., Bianco, A., & Macchiarini, P., (2014). Electrospun gelatin scaffolds incorporating rat decellularized brain extracellular matrix for neural tissue engineering. *Biomaterials, 35*(4), 1205–1214.

Bigi, A., Cojazzi, G., Panzavolta, S., Rubini, K., & Roveri, N., (2001). Mechanical and thermal properties of gelatin films at different degrees of glutaraldehyde crosslinking. *Biomaterials, 22*(8), 763–768.

Burke, C. J., Hsu, T. A., & Volkin, D. B., (1999). Formulation, stability, and delivery of live attenuated vaccines for human use. *Critical Reviews™ in Therapeutic Drug Carrier Systems, 16*(1), 1–83.

Chen, H., Huang, J., Yu, J., Liu, S., & Gu, P., (2011). Electrospun chitosan-graft-poly (ε-caprolactone)/poly (ε-caprolactone) cationic nanofibrous mats as potential scaffolds for skin tissue engineering. *International Journal of Biological Macromolecules, 48*(1), 13–19.

Chen, J., Liu, Z., Chen, M., Zhang, H., & Li, X., (2016). Electrospun gelatin fibers with a multiple release of antibiotics accelerate dermal regeneration in infected deep burns. *Macromolecular Bioscience, 16*(9), 1368–1380.

Chen, Y. C., Lin, R. Z., Qi, H., Yang, Y., Bae, H., Melero-Martin, J. M., & Khademhosseini, A., (2012). Functional human vascular network generated in photocrosslinkable gelatin methacrylate hydrogels. *Advanced Functional Materials, 22*(10), 2027–2039.

Cheow, C. S., Norizah, M. S., Kyaw, Z. Y., & Howell, N. K., (2007). Preparation and characterization of gelatins from the skins of sin croaker (Johnius dussumieri) and shortfin scad (Decapterus macrosoma). *Food Chemistry, 101*(1), 386–391.

Ciobanu, B. C., Cadinoiu, A. N., Popa, M., Desbrieres, J., & Peptu, C. A., (2014). Modulated release from liposomes entrapped in chitosan/gelatin hydrogels. *Materials Science and Engineering: C, 43*, 383–391.

Cohen, B., Panker, M., Zuckerman, E., Foox, M., & Zilberman, M., (2014). Effect of calcium phosphate-based fillers on the structure and bonding strength of novel gelatin–alginate bioadhesives. *Journal of Biomaterials Applications, 28*(9), 1366–1375.

Cohen, B., Pinkas, O., Foox, M., & Zilberman, M., (2013). Gelatin–alginate novel tissue adhesives and their formulation–strength effects. *Acta Biomaterialia, 9*(11), 9004–9011.

Cohen, B., Shefy-Peleg, A., & Zilberman, M., (2014). Novel gelatin/alginate soft tissue adhesives loaded with drugs for pain management: Structure and properties. *Journal of Biomaterials Science, Polymer Edition, 25*(3), 224–240.

Cui, L., Jia, J., Guo, Y., Liu, Y., & Zhu, P., (2014). Preparation and characterization of IPN hydrogels composed of chitosan and gelatin cross-linked by genipin. *Carbohydrate Polymers, 99*, 31–38.

De Almeida, P. F., Da Silva, J. R., Da Silva, L. S. C., De Brito, F. T. M., & Santana, J. C. C., (2013). Quality assurance and economical feasibility of an innovative product obtained from a by-product of the meat industry in Brazil. *African Journal of Business Management, 7*(27), 2410–2419.

Divya, M., Vaseeharan, B., Abinaya, M., Vijayakumar, S., Govindarajan, M., Alharbi, N. S., Kadaikunnan, S., Khaled, J. M., & Benelli, G., (2018). Biopolymer gelatin-coated zinc oxide nanoparticles showed high antibacterial, anti-biofilm and anti-angiogenic activity. *Journal of Photochemistry and Photobiology B: Biology, 178*, 211–218.

Feng, Q., Wei, K., Lin, S., Xu, Z., Sun, Y., Shi, P., Li, G., & Bian, L., (2016). Mechanically resilient, injectable, and bioadhesive supramolecular gelatin hydrogels crosslinked by weak host-guest interactions assist cell infiltration and *in situ* tissue regeneration. *Biomaterials, 101*, 217–228.

Foox, M., & Zilberman, M., (2015). Drug delivery from gelatin-based systems. *Expert Opinion on Drug Delivery, 12*(9), 1547–1563.

Gan, Z., Zhang, T., Liu, Y., & Wu, D., (2012). Temperature-triggered enzyme immobilization and release based on cross-linked gelatin nanoparticles. *PloS ONE, 7*(10), p.e47154.

Guan, P., Lu, Y., Qi, J., Niu, M., Lian, R., & Wu, W., (2015). Solidification of liposomes by freeze-drying: The importance of incorporating gelatin as interior support on enhanced physical stability. *International Journal of Pharmaceutics, 478*(2), 655–664.

Haddar, A., Sellimi, S., Ghannouchi, R., Alvarez, O. M., Nasri, M., & Bougatef, A., (2012). Functional, antioxidant and film-forming properties of tuna-skin gelatin with a brown algae extract. *International Journal of Biological Macromolecules, 51*(4), 477–483.

Haroun, A. A., El-Halawany, N. R., Loira-Pastoriza, C., & Maincent, P., (2014). Synthesis and *in vitro* release study of ibuprofen-loaded gelatin graft copolymer nanoparticles. *Drug Development and Industrial Pharmacy, 40*(1), 61–65.

Hoare, T. R., & Kohane, D. S., (2008). Hydrogels in drug delivery: Progress and challenges. *Polymer, 49*(8), 1993–2007.

Huang, K. S., Yang, C. H., Kung, C. P., Grumezescu, A. M., Ker, M. D., Lin, Y. S., & Wang, C. Y., (2014). Synthesis of uniform core-shell gelatin-alginate microparticles as intestine-released oral delivery drug carrier. *Electrophoresis, 35*(2/3), 330–336.

Hutson, C. B., Nichol, J. W., Aubin, H., Bae, H., Yamanlar, S., Al-Haque, S., Koshy, S. T., & Khademhosseini, A., (2011). Synthesis and characterization of tunable poly (ethylene glycol): Gelatin methacrylate composite hydrogels. *Tissue Engineering Part A, 17*(13/14), 1713–1723.

Jaipan, P., Nguyen, A., & Narayan, R. J., (2017). Gelatin-based hydrogels for biomedical applications. *MRS Communications, 7*(3), 416–426.

Jongjareonrak, A., Benjakul, S., Visessanguan, W., Nagai, T., & Tanaka, M., (2005). Isolation and characterization of acid and pepsin-solubilized collagens from the skin of Brownstripe red snapper (Lutjanus vitta). *Food Chemistry, 93*(3), 475–484.

Jongjareonrak, A., Rawdkuen, S., Chaijan, M., Benjakul, S., Osako, K., & Tanaka, M., (2010). Chemical compositions and characterization of skin gelatin from farmed giant catfish (Pangasianodon gigas). *LWT-Food Science and Technology, 43*(1), 161–165.

Joshy, K. S., Snigdha, S., Kalarikkal, N., Pothen, L. A., & Thomas, S., (2017). Gelatin modified lipid nanoparticles for anti-viral drug delivery. *Chemistry and Physics of Lipids, 207*, 24–37.

Jridi, M., Nasri, R., Salem, R. B. S. B., Lassoued, I., Barkia, A., Nasri, M., & Souissi, N., (2015). Chemical and biophysical properties of gelatins extracted from the skin of octopus (Octopus vulgaris). *LWT-Food Science and Technology, 60*(2), 881–889.

Jridi, M., Souissi, N., Mbarek, A., Chadeyron, G., Kammoun, M., & Nasri, M., (2013). Comparative study of physicomechanical and antioxidant properties of edible gelatin films from the skin of cuttlefish. *International Journal of Biological Macromolecules, 61*, 17–25.

Jun, J. Y., Nguyen, H. H., Chun, H. S., Kang, B. C., & Ko, S., (2011). Preparation of size-controlled bovine serum albumin (BSA) nanoparticles by a modified desolvation method. *Food Chemistry, 127*(4), 1892–1898.

Karim, A. A., & Bhat, R., (2009). Fish gelatin: Properties, challenges, and prospects as an alternative to mammalian gelatins. *Food Hydrocolloids, 23*(3), 563–576.

Kim, H. W., Yu, H. S., & Lee, H. H., (2008). Nanofibrous matrices of poly (lactic acid) and gelatin polymeric blends for the improvement of cellular responses. *Journal of Biomedical Materials Research Part A, 87*(1), 25–32.

Kittiphattanabawon, P., Benjakul, S., Visessanguan, W., & Shahidi, F., (2012). Gelatin hydrolysate from blacktip shark skin prepared using papaya latex enzyme: Antioxidant activity and its potential in model systems. *Food Chemistry, 135*(3), 1118–1126.

Lam, P. L., Gambari, R., Kok, S. L., Lam, K. H., Tang, J. O., Bian, Z. X., Lee, K. H., & Chui, C. H., (2015). Non-toxic agarose/gelatin-based microencapsulation system containing gallic acid for antifungal application. *International Journal of Molecular Medicine, 35*(2), 503–510.

Lankalapalli, S., & Vinai, K. T. V., (2016). Formulation and evaluation of rifampicin liposomes for buccal drug delivery. *Current Drug Delivery, 13*(7), 1084–1099.

Li, W. M., Su, C. W., Chen, Y. W., & Chen, S. Y., (2015). In situ DOX-calcium phosphate mineralized CPT-amphiphilic gelatin nanoparticle for intracellular controlled sequential release of multiple drugs. *Acta Biomaterialia, 15*, 191–199.

Lin, R. Z., Chen, Y. C., Moreno-Luna, R., Khademhosseini, A., & Melero-Martin, J. M., (2013). Transdermal regulation of vascular network bioengineering using a photopolymerizable methacrylated gelatin hydrogel. *Biomaterials, 34*(28), 6785–6796.

Liu, X., Smith, L. A., Hu, J., & Ma, P. X., (2009). Biomimetic nanofibrous gelatin/apatite composite scaffolds for bone tissue engineering. *Biomaterials, 30*(12), 2252–2258.

Mad-Ali, S., Benjakul, S., Prodpran, T., & Maqsood, S., (2016). Interfacial properties of gelatin from goat skin as influenced by drying methods. *LWT-Food Science and Technology, 73*, 102–107.

Madhamuthanalli, C. V., & Bangalore, S. A., (2014). Rheological and physicochemical properties of gelatin extracted from the skin of a few species of freshwater carp. *International Journal of Food Science & Technology, 49*(7), 1758–1764.

Mahor, A., Prajapati, S. K., Verma, A., Gupta, R., Iyer, A. K., & Kesharwani, P., (2016). Moxifloxacin loaded gelatin nanoparticles for ocular delivery: Formulation and *in-vitro, in-vivo* evaluation. *Journal of Colloid and Interface Science, 483*, 132–138.

Maleknia, L., & Majdi, Z. R., (2014). Electrospinning of gelatin nanofiber for biomedical application. *Oriental Journal of Chemistry, 30*(4), 2043–2048.

Manca, M. L., Cassano, R., Valenti, D., Trombino, S., Ferrarelli, T., Picci, N., Fadda, A. M., & Manconi, M., (2013). Isoniazid-gelatin conjugate microparticles containing rifampicin for the treatment of tuberculosis. *Journal of Pharmacy and Pharmacology, 65*(9), 1302–1311.

Moraru, A. D., Costuleanu, M., Sava, A., Costin, D., Peptu, C., Popa, M., & Chiseliţă, D., (2014). Intraocular biodistribution of intravitreal injected chitosan/gelatin nanoparticles. *Rom. J. Morphol. Embryol, 55*(3), 869–875.

Nayak, D., Boxi, A., Ashe, S., Thathapudi, N. C., & Nayak, B., (2017). Stavudine loaded gelatin liposomes for HIV therapy: Preparation, characterization and *in vitro* cytotoxic evaluation. *Materials Science and Engineering: C, 73*, 406–416.

Nejat, H., Rabiee, M., Varshochian, R., Tahriri, M., Jazayeri, H. E., Rajadas, J., Ye, H., Cui, Z., & Tayebi, L., (2017). Preparation and characterization of cardamom extract-loaded gelatin nanoparticles as effective targeted drug delivery system to treat glioblastoma. *Reactive and Functional Polymers, 120*, 46–56.

Nguyen, A. H., McKinney, J., Miller, T., Bongiorno, T., & McDevitt, T. C., (2015). Gelatin methacrylate microspheres for controlled growth factor release. *Acta Biomaterialia, 13*, 101–110.

Nichol, J. W., Koshy, S. T., Bae, H., Hwang, C. M., Yamanlar, S., & Khademhosseini, A., (2010). Cell-laden micro-engineered gelatin methacrylate hydrogels. *Biomaterials, 31*(21), 5536–5544.

Nunes, P. S., Rabelo, A. S., De Souza, J. C. C., Santana, B. V., Da Silva, T. M. M., Serafini, M. R., et al., (2016). Gelatin-based membrane containing usnic acid-loaded liposome improves dermal burn healing in a porcine model. *International Journal of Pharmaceutics, 513*(1), 473–482.

Ofokansi, K., Winter, G., Fricker, G., & Coester, C., (2010). Matrix-loaded biodegradable gelatin nanoparticles as new approach to improve drug loading and delivery. *European Journal of Pharmaceutics and Biopharmaceutics*, *76*(1), 1–9.

Olsen, D., Yang, C., Bodo, M., Chang, R., Leigh, S., Baez, J., et al., (2003). Recombinant collagen and gelatin for drug delivery. *Advanced Drug Delivery Reviews*, *55*(12), 1547–1567.

Otoni, C. G., Avena-Bustillos, R. J., Chiou, B. S., Bilbao-Sainz, C., Bechtel, P. J., & McHugh, T. H., (2012). Ultraviolet-B radiation induced cross-linking improves physical properties of cold-and warm-water fish gelatin gels and films. *Journal of Food Science*, *77*(9) E215–E223.

Pierce, B. F., Pittermann, E., Ma, N., Gebauer, T., Neffe, A. T., Hölscher, M., Jung, F., & Lendlein, A., (2012). Viability of human mesenchymal stem cells seeded on crosslinked entropy-elastic gelatin-based hydrogels. *Macromolecular Bioscience*, *12*(3), 312–321.

Pinkas, O., & Zilberman, M., (2014). Effect of hemostatic agents on properties of gelatin–alginate soft tissue adhesives. *Journal of Biomaterials Science, Polymer Edition*, *25*(6), 555–573.

Rajan, M., & Raj, V., (2013). Formation and characterization of chitosan-polylactic acid-polyethylene glycol-gelatin nanoparticles: A novel biosystem for controlled drug delivery. *Carbohydrate Polymers*, *98*(1), 951–958.

Rajzer, I., Menaszek, E., Kwiatkowski, R., Planell, J. A., & Castano, O., (2014). Electrospun gelatin/poly (ε-caprolactone) fibrous scaffold modified with calcium phosphate for bone tissue engineering. *Materials Science and Engineering: C*, *44*, 183–190.

Ratanavaraporn, J., Rangkupan, R., Jeeratawatchai, H., Kanokpanont, S., & Damrongsakkul, S., (2010). Influences of physical and chemical crosslinking techniques on electrospun type A and B gelatin fiber mats. *International Journal of Biological Macromolecules*, *47*(4), 431–438.

Rault, I., Frei, V., Herbage, D., Abdul-Malak, N., & Huc, A., (1996). Evaluation of different chemical methods for cross-linking collagen gel, films and sponges. *Journal of Materials Science: Materials in Medicine*, *7*(4), 215–221.

Sahraee, S., Milani, J. M., Ghanbarzadeh, B., & Hamishehkar, H., (2017). Physicochemical and antifungal properties of bio-nanocomposite film based on gelatin-chitin nanoparticles. *International Journal of Biological Macromolecules*, *97*, 373–381.

Sakaguchi, M., Toda, M., Ebihara, T., Irie, S., Hori, H., Imai, A., Yanagida, M., Miyazawa, H., Ohsuna, H., Ikezawa, Z., & Inouye, S., (2000). IgE antibody to fish gelatin (type I collagen) in patients with fish allergy. *Journal of Allergy and Clinical Immunology*, *106*(3), 579–584.

Sarbon, N. M., Badii, F., & Howell, N. K., (2013). Preparation and characterization of chicken skin gelatin as an alternative to mammalian gelatin. *Food Hydrocolloids*, *30*(1), 143–151.

Shende, P., & Gaud, R., (2009). Formulation and comparative characterization of chitosan, gelatin, and chitosan–gelatin-coated liposomes of CPT-11–HCl. *Drug Development and Industrial Pharmacy*, *35*(5), 612–618.

Shutava, T. G., Balkundi, S. S., Vangala, P., Steffan, J. J., Bigelow, R. L., Cardelli, J. A., O'Neal, D. P., & Lvov, Y. M., (2009). Layer-by-layer-coated gelatin nanoparticles as a vehicle for delivery of natural polyphenols. *ACS Nano*, *3*(7), 1877–1885.

Singh, A., Xu, J., Mattheolabakis, G., & Amiji, M., (2016). EGFR-targeted gelatin nanoparticles for systemic administration of gemcitabine in an orthotopic pancreatic cancer model. *Nanomedicine: Nanotechnology, Biology and Medicine, 12*(3), 589–600.

Singh, V., & Ahmad, S., (2014). Carboxymethyl cellulose-gelatin-silica nanohybrid: An efficient carrier matrix for alpha-amylase. *International Journal of Biological Macromolecules, 67,* 439–445.

Singh, V., & Ahmed, S., (2012). Silver nanoparticle (AgNPs) doped gum acacia–gelatin–silica nanohybrid: An effective support for diastase immobilization. *International Journal of Biological Macromolecules, 50*(2), 353–361.

Sivera, M., Kvitek, L., Soukupova, J., Panacek, A., Prucek, R., Vecerova, R., & Zboril, R., (2014). Silver nanoparticles modified by gelatin with extraordinary pH stability and long-term antibacterial activity. *PloS ONE, 9*(8), p.e103675.

Tabarestani, H. S., Maghsoudlou, Y., Motamedzadegan, A., & Mahoonak, A. S., (2010). Optimization of physicochemical properties of gelatin extracted from fish skin of rainbow trout (Oncorhynchus mykiss). *Bioresource Technology, 101*(15), 6207–6214.

Tonsomboon, K., & Oyen, M. L., (2013). Composite electrospun gelatin fiber-alginate gel scaffolds for mechanically robust tissue-engineered cornea. *Journal of the Mechanical Behavior of Biomedical Materials, 21,* 185–194.

Tseng, C. L., Su, W. Y., Yen, K. C., Yang, K. C., & Lin, F. H., (2009). The use of biotinylated-EGF-modified gelatin nanoparticle carrier to enhance cisplatin accumulation in cancerous lungs via inhalation. *Biomaterials, 30*(20), 3476–3485.

Tu, Z. C., Huang, T., Wang, H., Sha, X. M., Shi, Y., Huang, X. Q., Man, Z. Z., & Li, D. J., (2015). Physico-chemical properties of gelatin from bighead carp (Hypophthalmichthys nobilis) scales by ultrasound-assisted extraction. *Journal of Food Science and Technology, 52*(4), 2166–2174.

Vijayaraghavan, R., Thompson, B. C., MacFarlane, D. R., Kumar, R., Surianarayanan, M., Aishwarya, S., & Sehgal, P. K., (2009). Biocompatibility of choline salts as crosslinking agents for collagen-based biomaterials. *Chemical Communications, 46*(2), 294–296.

Wang, H., Feng, Y., Fang, Z., Xiao, R., Yuan, W., & Khan, M., (2013). Fabrication and characterization of electrospun gelatin-heparin nanofibers as vascular tissue engineering. *Macromolecular Research, 21*(8), 860–869.

Wang, T., Zhu, X. K., Xue, X. T., & Wu, D. Y., (2012). Hydrogel sheets of chitosan, honey and gelatin as burn wound dressings. *Carbohydrate Polymers, 88*(1), 75–83.

Wise, D. L., (2000). *Handbook of Pharmaceutical Controlled Release Technology.* CRC Press.

Xu, M., Wei, L., Xiao, Y., Bi, H., Yang, H., & Du, Y., (2017). Physicochemical and functional properties of gelatin extracted from Yak skin. *International Journal of Biological Macromolecules, 95,* 1246–1253.

Xue, J., Feng, B., Zheng, R., Lu, Y., Zhou, G., Liu, W., Cao, Y., Zhang, Y., & Zhang, W. J., (2013). Engineering ear-shaped cartilage using electrospun fibrous membranes of gelatin/polycaprolactone. *Biomaterials, 34*(11), 2624–2631.

Zhao, Y. Z., Li, X., Lu, C. T., Xu, Y. Y., Lv, H. F., Dai, D. D., et al., (2012b). Experiment on the feasibility of using modified gelatin nanoparticles as insulin pulmonary administration system for diabetes therapy. *Acta Diabetologica, 49*(4), 315–325.

Zhao, Z. X., Gao, S. Y., Wang, J. C., Chen, C. J., Zhao, E. Y., Hou, W. J., et al., (2012a). Self-assembly nanomicelles based on cationic mPEG-PLA-b-polyarginine (R 15) triblock copolymer for siRNA delivery. *Biomaterials*, *33*(28), 6793–6807.

Zillies, J. C., Zwiorek, K., Hoffmann, F., Vollmar, A., Anchordoquy, T. J., Winter, G., & Coester, C., (2008). Formulation development of freeze-dried oligonucleotide-loaded gelatin nanoparticles. *European Journal of Pharmaceutics and Biopharmaceutics*, *70*(2), 514–521.

Zorzi, G. K., Contreras-Ruiz, L., Párraga, J. E., López-García, A., Romero, B. R., Diebold, Y., Seijo, B., & Sánchez, A., (2011b). Expression of MUC5AC in ocular surface epithelial cells using cationized gelatin nanoparticles. *Molecular Pharmaceutics, 8*(5), 1783–1788.

Zorzi, G. K., Párraga, J. E., Seijo, B., & Sánchez, A., (2011a). Hybrid nanoparticle design based on cationized gelatin and the polyanions dextran sulfate and chondroitin sulfate for ocular gene therapy. *Macromolecular Bioscience*, *11*(7), 905–913.

Zou, T., Percival, S. S., Cheng, Q., Li, Z., Rowe, C. A., & Gu, L., (2012). Preparation, characterization, and induction of cell apoptosis of cocoa procyanidins–gelatin–chitosan nanoparticles. *European Journal of Pharmaceutics and Biopharmaceutics*, *82*(1), 36–42.

Zwiorek, K., Bourquin, C., Battiany, J., Winter, G., Endres, S., Hartmann, G., & Coester, C., (2008). Delivery by cationic gelatin nanoparticles strongly increases the immunostimulatory effects of CpG oligonucleotides. *Pharmaceutical Research*, *25*(3), 551–562.

CHAPTER 5

Pharmaceutical Applications of Chondroitin

DILIPKUMAR PAL,[1] AMIT KUMAR NAYAK,[2] SUPRIYO SAHA,[3] and MD SAQUIB HASNAIN[4]

[1]Department of Pharmaceutical Sciences, Guru Ghasidas Vishwavidyalaya, Koni, Bilaspur–495009, C.G., India

[2]Department of Pharmaceutics, Seemanta Institute of Pharmaceutical Sciences, Mayurbhanj–757086, Odisha, India

[3]School of Pharmaceutical Sciences and Technology, Sardar Bhagwan Singh University, Dehradun – 248161, Uttrakhand, India

[4]Department of Pharmacy, Shri Venkateshwara University, Gajraula, U.P., India

ABSTRACT

Chondroitin is mostly available as CS form, which is a sulfated glycosaminoglycan comprising D-glucuronic residue and *N*-Acetyl D-galactosamine residue. In supplements, CS is usually isolated from cartilages of shark, cow, and pig. Despite of its greater variability of the source of extraction, the activity profile of CS was observed. Because of the biodegradable and biocompatible nature, it is the primary choice to treat osteoarthritis and tissue engineering applications. CS has been used for pharmaceutical applications, mainly in drug delivery and drug targeting. CS has also been used for controlled release drug carrier, drug targeting (for anticancer drugs), protein delivery, etc. The chapter provides a detailed overview of pharmaceutical uses of CS obtained from the natural source.

5.1 INTRODUCTION

Because of the recent developments in the use of natural products, researchers are recently tending towards the replacement of synthetically derived polymers by various naturally derived polymers owing to their ready availability from the natural resources, very low extraction expenditure, biodegradability, biocompatibility, etc. (Hasnain et al., 2018a, b; Nayak et al., 2012, 2015, 2018a, b). At present, huge numbers of naturally derived polymers are available and exploited as the pharmaceutical excipients (Jena et al., 2018; Nayak and Pal, 2015, 2017a, b). The uses of natural polymers in the pharmaceutical industry play very important roles as the natural sources are recognized as potential renewable as well as sustainable sources offering constant supplies of these materials (Guru et al., 2018; Hasnain and Nayak, 2018a; Pal and Nayak, 2015a, b, 2017). Most of the naturally derived polymers are extracted from plant parts (Nayak, 2016; Nayak et al., 2018c; Nayak and Pal, 2015, 2018), algae (Hasnain and Nayak, 2018b), animals (Ikrima et al., 2018; Xu et al., 2017), etc. Almost all these naturally derived polymers are either carbohydrate or proteins in nature (Nayak and Pal, 2016a, b; Nayak et al., 2016). Recent years, numerous protein-based natural polymers have emerged as an attractive category of naturally derived polymers because of their advantageous physicochemical, chemical, as well as biological properties (Hasnain et al., 2010). The important biological properties of these protein-based natural polymers are biocompatibility and biodegradation (Sleep, 2014). These are usually composed of repeated units of amino acids. In addition, the molecular weights and sequences of amino acids of these protein-based natural polymers can be controlled specifically, which finally verifies the biodegradability, pharmacokinetic outcomes and biological activity (Sleep, 2014; Rajan and Raj, 2016).

Chondroitin sulfate (CS) is an important protein-based natural polymer occurred in the human connective tissues especially found in cartilage and bone, which provides a resistance towards compression and also is occurred in the hyaline cartilage and synovial fluid (Sherman et al., 2012). In natural supplements, CS is usually obtained from animal cartilages. CS is chemically a sulfated glycosaminoglycan, which is generally occurred as attached to the protein(s) of the proteoglycan structure (Dyck and Abdolrezaee, 2015). Chondroitin is slowly absorbed from the gastrointestinal tract. CS possesses the unbranched polysaccharide structure of the various polymeric chain lengths contained two alternating monosaccharide residues: D-glucuronic

Pharmaceutical Applications of Chondroitin 119

residue and *N*-Acetyl D-galactosamine residue (Sherman et al., 2012). CS chains are attached with the hydroxyl groups on the serine residues of proteins. The functioning of chondroitin depends upon the properties of the attached proteoglycan. The sulfate groups of CS is condencedly packed and greatly charged that produce an electrostatical repulsion facilitating the resistance of cartilage to the compression (Miyata and Kitagawa, 2017). Chondroitin interacts with proteins in the extracellular matrix (ECM), regulating a diverse array of the cellular functions. Due to its anti-inflammatory and immunomodulatory activity, CS shows a good effect on osteoarthritis. CS is capable of inducing the synthesis of hyaluronic acid (HA) and proteoglycans. Also, it inhibits the synthesis of proteolytic enzymes and nitric oxide (Kwok et al., 2012; Silver and Silver, 2014).

5.2 STRUCTURE OF CS

CS is a double sugar moiety contained with D-glucuronic residue and *N*-Acetyl D-galactosamine residue, which are linked by the glycosidic linkage within D-glucuronic Acid within C1 to C3 and C1 to C4 N-acetyl D-galactosamine to develop an unbranched glycosa aminoglycan chains (Miller and Clegg, 2011; Miyata and Kitagawa, 2017). One single glycosaminoglycan chain is contained maximum to 50 double sugar moieties. Modification in the glycosaminoglycans is done by sulfation mechanism, and it is profound at C-4 or C-6 of N-acetyl D-galactosamine, C-2 in the iduronic acid and occasionally in the C-3 of D-glucuronic acid (Miller and Clegg, 2011).

5.3 SYNTHESIS OF CS AND ITS MODIFICATIONS

Although the basic unit of chondroitin sulfate is double sugar moiety; which formed as growing glycosaminoglycan chain with alternate addition of N-acetyl galactosamine and D-glucuronic acid (Miyata and Kitagawa, 2017). Chondroitin synthases, N-acetyl galactosamine transferase, D-glucuronic acid transferases, CS N-acetyl galactosaminyl transferase I & II, chondroitin sulfate glucuronyltransferase and CS polymerization factor catalyze the addition of N-acetyl galactosamine and D-glucuronic acid with proper coordination during chain elongation by phosphorylation or sulfation reaction, but not a single enzyme can develop the total chondroitin sulfate chain. Phosphorylation reaction on the Xyl residue present in the tetrasaccharide portion not yet recognized the required detailed analysis of the role of sulfation reaction within the CS chains (Figure 5.1). Enzymes of the carbohydrate sulfotransferase family are responsible for the sulfation modification with Chondroitin 4-sulfotransferase, chondroitin 6-sulfotransferase, and GalNAc-4 sulfate 6-O-sulfotransferase (Kwok et al., 2012).

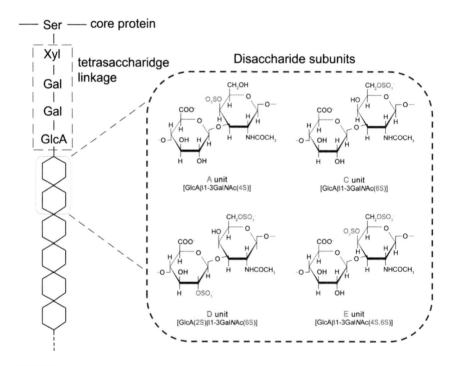

FIGURE 5.1 (See color insert.) A schematic diagram showing the structure of CS. CS-glycosaminoglycans are attached to the serine residue on the core protein via a tetrasaccharide linkage.
Source: Kwok et al., (2012); Copyright © 2012 with permission from Elsevier B.V.

5.4 PHARMACEUTICAL APPLICATION OF CHONDROITIN

5.4.1 LOXOPROFEN HYDROGELS COMPOSED WITH CS

Mixtures of CS and AMPS (i.e., 2-acrylamido-2-methyl-1-propane sulfonic acid) monomer in the presence of N, N'-methylene bisacrylamide

Pharmaceutical Applications of Chondroitin   121

were used to form hydrogels with free radical copolymerization property (Ikrima et al., 2018). Scanning electron microscopy (SEM), Fourier transform infrared (FTIR), X-ray diffraction (XRD), thermogravimetry analysis (TGA), and differential scanning calorimetry (DSC) data of hydrogel were showed the proper structure of formulation. SEM results confirmed the topology of hydrogels with highly porous modified swelling characteristics. FTIR results were revealed the reaction in-between CS and N, N'-methylene bisacrylamide and convinced the incorporation of N, N'-methylene bisacrylamide chains into data backbone. TGA results revealed thermo-resistant crosslinking nature. The effects of pH confirmed that swelling properties were increased with a higher concentration of CS or AMPS. Also, the loading efficacies of the drug was examined the distribution of loxoprofen sodium with swelling characteristics. Drug release profiles (*in vitro*) and kinetic studies of formulation confirmed the reproducible and reliable results. Release kinetics from hydrogels was correlated with the Higuchi model, indicated proper released behavior. Also, the Korsmeyer–Peppas model observed greater diffusion in the hydrated matrix and polymer relaxation (Figure 5.2). Furthermore, AMPS hydrogels behave as a controlled drug releasing system (Ikrima et al., 2018).

5.4.2 SELF-ASSEMBLED CS-NISIN NANOGEL

Cationic nisin and anionic CS as biological macromolecule were electrostatically complexed to form a self-assembled nanogel (Mohtashamiana et al., 2018). The different characteristics of chondroitin sulfate-nisin nanogel were identified by the Central Composite design and Placket-Burman design. The ratio of nisin and chondroitin, rate of nisin solution addition and magnetic stirrer, type, buffer pH and temperature and rotation of centrifuge cooling were analyzed as the independent factors to assess the CS-nisin nanogel. The drug entrapment efficiency was statistically regulated by the concentration ratio. Also, the nisin injection rate and buffer pH were regulated by the hydrodynamic diameter as well as the loading capacity of the nanogel. Nisin to CS ratio, pH, and injection rate were identified as the principal factor to develop an optimized nanogel structure. The response surface methodology (RSM) model was employed for the evaluation of the closeness in-between observed and predicted values.

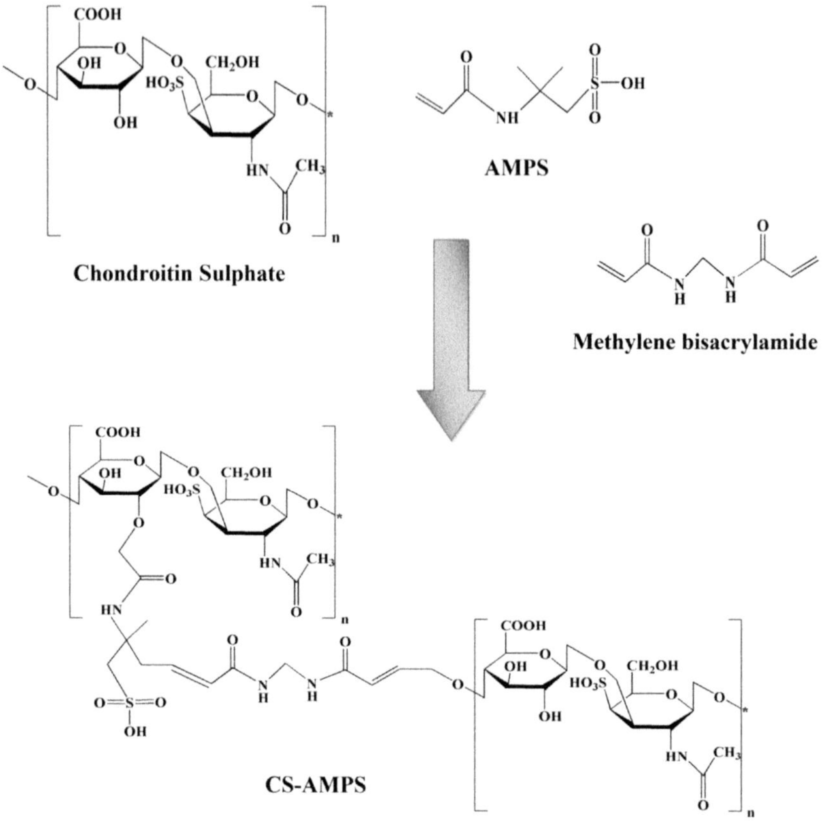

FIGURE 5.2 Formation of CS-AMPS (2-acrylamido-2-methyl-1-propane sulfonic acid) hydrogel.

Source: Khalid et al., (2018) Copyright © 2017 with permission from Elsevier B.V.

5.4.3 CS-BASED NANOGELS CONTAINING DOXORUBICIN HYDROCHLORIDE

Inverse microemulsion polymerization technique was used to develop chondroitin nanogel (Juqun et al., 2012). The properties of formulated CS-based nanogels were investigated by light scatter technique, FTIR, ^{1}H nuclear magnetic resonance (^{1}H NMR) and transmission electron microscopy (TEM). Size of nanogel was 145–340 nm as revealed by outcomes with the variable substitution of maleoyl in CS, and pH of the formulation showed the greater stability in the aqueous milieu, and the MTT assay results revealed lesser cell toxicity, *in vitro*. When the doxorubicin-loaded nanogel showed that the high loading ability and drug release were dependent upon pH and porosity of the formulation. Thus, this nanogel was found biologically compatible and degradable with greater drug delivery (Juqun et al., 2012).

5.4.4   CONJUGATION OF CHONDROITIN SULFATE A (CS A) AND DEOXYCHOLIC ACID FOR TRIGGERED RELEASING OF DOXORUBICIN

Reduction-sensitive conjugation of chondroitin sulfate A (CS A) and deoxycholic acid using disulfide bonding was developed. This biologically reduced formulation was formed micelles in the aqueous environment (Hongxia et al., 2017). Critical micelle concentration of CS A-ss- deoxycholic acid conjugate was showed 0.047 mg/mL of critical micelle concentration with 387 nm of mean diameter. Drug loading capacity was done by encapsulated doxorubicin and release behavior was examined in 7.4 phosphates buffered saline; whereas the similar release characteristics were observed with the reduction-sensitive micelles and reduction-insensitive control micelles. *In vitro* releasing of doxorubicin from the responsive micelles was accelerated in 20 mM glutathione-phosphate buffer saline. Furthermore, the confocal laser scanning microscopy was confirmed that doxorubicin-loaded CS A-ss-deoxycholic acid conjugate micelles were observed with greater efficiency against human gastric cancer cell line (HGC-27). The results of this research revealed that reduction sensitive doxorubicin-loaded CS A-ss-deoxycholic acid conjugate micelles behaved as potent intracellular carriers of anticancer drugs (Hongxia et al., 2017).

5.4.5   POLYETHYLENE GLYCOL-CS A NANOPARTICLES FOR TUMOR-TARGETED DELIVERY

Polyethylene glycol-decorated CS A-deoxycholic acid nanoparticles were developed by the reaction of amino group polyethylene glycol and free acid group of CS A-deoxycholic acid nanoparticles with the mean diameter of 247 nm, negative zeta potential, greater than 90% of drug loading of doxorubicin-loaded nanoparticle for patients with the ovary cancer (Jae-Young et al., 2016). Sustained and pH-dependent doxorubicin releasing profiles from polyethylene glycol-decorated CS A-deoxycholic acid nanoparticles were observed in the dissolution testing, which showed the pH-dependent and sustained releasing of doxorubicin. Flow properties and confocal laser scanning studies were used to evaluate the endocytosis of nanoparticles by SKOV-3 cells based on the interaction with the CS A-CD44 receptor. Therefore, these developed polyethylene glycol-decorated CS A-deoxycholic acid nanoparticles can work as a promising delivery vehicle for ovarian cancers.

5.4.6 BILE ACID CONJUGATED CS A-BASED NANOPARTICLES FOR ANTICANCER DRUG DELIVERY

Hydrophilic CS A and hydrophobic deoxycholic acid-based nanoparticles were produced via amide bond formation with a mean diameter of 230 nm, small distribution of size, negatively zeta potential, and comparatively high encapsulation efficiency of drug (Jae-Young et al., 2015). These bile acid conjugated CS A-based nanoparticles demonstrated a sustained as well as pH-responsive drug releasing pattern for the tumor-targeting delivery of an anticancer drug, doxorubicin. Reaction in-between CS A and the CD44 receptor was reduced the cell uptake of nanoparticles in the human breast adenocarcinoma cell lines (MDA-MB-231). Therefore, these nanoparticles were used as a platform for CD44 receptor-positive cancers.

5.4.7 CS-COUPLED LIPOSOMES FOR TARGETING SOLID TUMOR

CS-coupled liposomes were developed to target solid tumor. Formulations were prepared using CS by film casting method (Bagari et al., 2011). The CS-coupled liposomes were evaluated by IR spectroscopy and other parameters like size, topology, surface potential, drug entrapment capacity, and drug release of vesicle were evaluated. At 256 nm, size of uncoupled liposome was less than CS-coupled liposomes at 310 nm. After 24 hours, *in vitro* drug releasing uncoupled liposomes demonstrated as a release of 44.20%; whereas CS-coupled liposomes demonstrated 38.30% of drug release. Fluorescence microscopy was used to examine the presence of CS-coupled as well as uncoupled liposomes by the breast cancer cell lines (MDA-MB-231). Liposomes increased the uptake of drug in a solid tumor, followed by the administration of CS-coupled liposomal

Pharmaceutical Applications of Chondroitin 125

formulations in comparison with that of the uncoupled liposomal formulations or free drug and employed as vectors for the targeting of solid tumors (Bagari et al., 2011).

5.4.8 CURCUMIN-LOADED IN CHITOSAN/CS NANOPARTICLES

Ionic gelation method was used to develop curcumin-loaded chitosan/CS nanoparticles with a polydispersity index of 0.151 ± 0.03 to 0.563 ± 0.07 and hydrodynamic diameter of 175.7 ± 2.5 to 710.2 ± 8.9 nm. The curcumin encapsulation of $62.4 \pm 0.61\%$ to $68.3 \pm 0.88\%$ in chitosan/CS nanoparticles and pH of the chitosan solution was the main factor (Katiuscia et al., 2015). At pH 6.8, the curcumin releasing from nanoparticles was done by a diffusion mechanism. MTT assay was showed significant reduction of 41.10 and 60.40% in the viability of A549 cells in the presence of both the curcumin-loaded and unloaded chitosan/CS nanoparticles.

5.4.9 COVALENT AND INJECTABLE CHITOSAN-CS HYDROGELS FOR DRUG DELIVERY

Natural polysaccharides were used to develop an injectable hydrogel and microsphere using Schiff based polysaccharide based covalent hydrogel. The cross-linking mechanism was used between the oxidized form of CS and carboxymethyl chitosan to form hydrogels (Ming et al., 2017). Cross-linking was formed by the reaction between $-NH_2$ and -CHO groups of the polymers. Also, bovine serum albumin (BSA) embedded chitosan microspheres (3.8–61.6) μm diameter were prepared by the emulsifying crosslinking method, later embedded into the hydrogels of carboxymethyl chitosan-oxidized CS to develop chitosan-based microsphere. Rate of the gel formation, topology, withstand properties, swelling property, degradation (*in vitro*) and BSA release of chitosan-based microspheres were verified. The outcomes attributed that withstand behavior, and biological property of formulations were statistically improved by embedding microspheres/hydrogels made of chitosan. Approximately, 30% of cumulative release was observed during 2 weeks of BSA from chitosan-based microspheres/gel embedded hydrogel. Furthermore, smaller swelling properties and a slower rate of degradation were observed from composite chitosan-based microsphere/hydrogel scaffolds than the control hydrogel without chitosan-based microspheres. *In vitro* encapsulation of the bovine articular chondrocytes within the chitosan-CS hydrogels was found as the potential injectable composite hydrogel for the cell delivery in cartilage tissue engineering (Ming et al., 2017).

5.4.10 CS-GOLD NANOPARTICLES FOR ORAL APPLICATION OF INSULIN

CS-gold nanoparticles were developed and evaluated its applicability for the oral delivery of insulin by using CS (Hyun-Jong et al., 2014). Insulin-loaded CS-gold nanoparticles having a mean diameter of 123 nm with a narrow particle size distribution and negative zeta potential were prepared. Surface plasmon resonance measurement study was used to confirm the selection of 0.50% w/v CS for the gold nanoparticle synthesis. Surface plasmon resonance study confirmed the stability of gold nanoparticles and gold nanoparticles/INS for 7 weeks. The cytotoxic effect on CaCo-2 cells, gold nanoparticles containing insulin was observed no significant cytotoxicity. Inducing of diabetes in the streptozotocin rat was confirmed that oral application of CS-gold nanoparticles containing insulin showed efficiently regulated blood glucose level as compared to that of the insulin-treated group. After 120 min of an oral dose of CS-gold nanoparticles containing insulin, mean insulin concentration in plasma was measured to have 6.61-fold greater than that of the insulin-treated group. All of these outcomes indicated that CS-gold nanoparticles containing insulin were successfully applied for the oral administration of insulin (Hyun-Jong et al., 2014).

5.4.11 CS-BASED POLYELECTROLYTE NANOPLEXES LOADED WITH CALCITONIN

CS based nanoparticles made of polyelectrolyte nanoplexes of CS/chitosan, CS/calcitonin, and CS/chitosan/calcitonin were formulated (Umerska et al., 2017). The properties CS/chitosan, CS/calcitonin, and CS/chitosan/calcitonin nanoparticles were also evaluated. The ratio of polymer, concentration, and molecular weight of chitosan were the regulating factors of both positively and negatively charged CS/chitosan nanoparticles. Calcitonin loaded CS/chitosan nanoparticles demonstrated 33% of loading capacity. Nanoparticles composed of CS and calcitonin (a binary system) exhibited

73% of drug loading capacity. Phosphate buffer solution, acetate buffer, and hydrochloric acid solutions helped to increase the particle size of CS-based nanoparticles as compared to that of the water. Even after 24 h, most of the particles were within the nanometer range. The media composition of the nanocarriers was affected by the calcitonin release (Umerska et al., 2017).

5.4.12 GELATIN-CS HYDROGEL WITH ANTIBACTERIAL EFFICACY

A new controlled release drug hydrogel was developed using a chemical crosslinking reaction of gelatin and CS for small cationic proteins (Alma et al., 2000). In the gelatin hydrogels, the amount of CS was 0–20% w/w. The hydrogels were characterized by density, swelling, and rheological behavior. The release of lysozyme embedded hydrogels was evaluated in the phosphate buffer saline system. The loading of lysozyme into the hydrogels was increased with the increasing of CS amounts in the formulations. The releasing rate of lysozyme was measured as 5–10% w/w decreased for the CS hydrogels, but the greater releasing rate was observed for the hydrogels with 20% w/w of CS. The two compartment diffusion cell was used to quantify the permeation of lysozyme through the gelatin-CS gels. Fluorescence recovery with photobleaching technique was used to evaluate the effective diffusion of lysozyme. These outcomes stated that minimal 5% w/w of CS was needed for the effective release (Alma et al., 2000).

5.4.13 CS-CHITOSAN MICROSPHERE AS CARRIER FOR PROTEIN DELIVERY

A macromolecule was developed using CS/chitosan microspheres and ovalbumin (model protein) by the emulsion-complex coacervation method (Katia et al., 2009). The structural integrity and physicochemical parameters were characterized by infrared spectroscopy, DSC, TGA, and XRD analyses. The degradation of the drug delivery system and the releasing of protein under the simulating conditions using intestinal fluids were observed by *in vitro* process. Chondroitinase ABC enzyme was used to evaluate the ability of bacterial enzymes produced by colonic microflora to degrade the systems. The outcomes revealed that different CS-chitosan composites were affected by both the stability of microparticles and protein releasing rate. The microspheres with 1:1 of CS-chitosan ratio was observed with a suitable ovalbumin release profile as the colon targeting therapy. In 24 hours, 30% of ovalbumin was released from the microspheres in various aqueous media as tested, whether 100% of protein was released in the presence of chondroitinase enzyme. Therefore, CS-chitosan microspheres can be used as a good protein delivery system for oral administration (Katia et al., 2009).

5.5 CONCLUSION

The chapter provides a detailed overview of recent pharmaceutical uses of chondroitin obtained from the natural sources. Chondroitin (a sulfated glycosaminoglycan) is mostly available as CS form. Due to the biodegradable and biocompatible nature, it is the primary choice to treat osteoarthritis and tissue engineering applications. CS has already been used in various pharmaceutical applications for controlled releasing, drug targeting (for anticancer drugs), protein delivery, etc. A variety of CS-based systems are already reported for showing the significant tumor-targeting applications. As a future scope, the calcitonin-like action of CS can be utilized with its pharmaceutical uses as an excipient, which can synergize the importance of active pharmaceutical ingredient.

KEYWORDS

- **chondroitin sulfate**
- **controlled release carrier**
- **drug delivery**
- **pharmaceutical excipient**

REFERENCES

Alma, J. K., Gerard, H. M. E., Tom, K. L. M., Stefaan, S. C. S., Joseph, D., Jeroen, K., Sebastian, A. J. Z., Jacob, D., & Jan, F., (2000). Combined gelatin-chondroitin sulfate hydrogels for controlled release of cationic antibacterial proteins. *Macromol., 33*, 3705–3713.

Bagari, R., Bansal, D., Gulbake, A., Jain, A., Soni, V., & Jain, S. K., (2011). Chondroitin sulfate functionalized liposomes for solid tumor targeting. *J. Drug. Targeting, 19*(4), 251–257.

Dyck, S. M., & Abdolrezaee, S. K., (2015). Chondroitin sulfate proteoglycans: Key modulators in the developing and pathologic central nervous system. *Exp. Neurol., 269*, 169–187.

Guru, P. R., Bera, H., Das, M., Hasnain, M. S., & Nayak, A. K., (2018). Aceclofenac-loaded *Plantago ovata* F. husk mucilage-Zn^{+2}-pectinate controlled-release matrices. *Starch – Stärke, 70*, 1700136.

Hasnain, M. S., & Nayak, A. K., (2018a). Alginate-inorganic composite particles as sustained drug delivery matrices. In: Inamuddin, A. A. M., & Mohammad, A., (eds.), *Applications of Nanocomposite Materials in Drug Delivery* (pp. 39–74). Elsevier Inc.

Hasnain, M. S., & Nayak, A. K., (2018b). Chitosan as responsive polymer for drug delivery applications. In: *Makhlouf, A. S. H., & Abu-Thabit, N. Y., (eds.), Stimuli-Responsive Polymeric Nanocarriers for Drug Delivery Applications* (Vol. 1, pp. 581–605). Types and triggers, Woodhead Publishing Series in Biomaterials, Elsevier Ltd.

Hasnain, M. S., Nayak, A. K., Singh, R., & Ahmad, F., (2010). Emerging trends of natural-based polymeric systems for drug delivery in tissue engineering applications. *Sci. J. UBU., 1*(2), 1–13.

Hasnain, M. S., Rishishwar, P., Rishishwar, S., Ali, S., & Nayak, A. K., (2018a). Extraction and characterization of cashew tree (*Anacardium occidentale*) gum, use in aceclofenac dental pastes. *Int. J. Biol. Macromol., 116*, 1074–1081.

Hasnain, M. S., Rishishwar, P., Rishishwar, S., Ali, S., & Nayak, A. K., (2018b). Isolation and characterization of *Linum usitatisimum* polysaccharide to prepare mucoadhesive beads of diclofenac sodium. *Int. J. Biol. Macromol., 116*, 162–172.

Hongxia, L., Shuqin, W., Jingmou, Y., Dun, F., Jin, R., Zhang, L., & Zhao, J., (2017). Reduction-sensitive micelles self-assembled from amphiphilic chondroitin sulfate A-deoxycholic acid conjugate for triggered release of doxorubicin. *Mater. Sci. Eng. C., 75*, 55–63.

Hyun-Jong, C., Jongsuk, O., Moon-Ki, C., Jong-In, H., Youmie, P., & Han-Joo, M., (2014). Chondroitin sulfate-capped gold nanoparticles for the oral delivery of insulin. *Int. J. Biol. Macromol., 63*, 15–20.

Jae-Young, L., Ju-Hwan, P., Jeong-Jun, L., Song, Y. L., Suk-Jae, C., Hyun-Jong, C., & Dae-Duk, K., (2016). Polyethylene glycol-conjugated chondroitin sulfate A derivative nanoparticles for tumor-targeted delivery of anticancer drugs. *Carbohydr Polym., 151*, 68–77.

Jae-Young, L., Suk-Jae, C., Hyun-Jong, C., & Dae-Duk, K., (2015). Bile acid-conjugated chondroitin sulfate A-based nanoparticles for tumor-targeted anticancer drug delivery. *Eur. J. Pharm. Biopharm., 94*, 532–541.

Jena, A. K., Nayak, A. K., De, A., Mitra, D., & Samanta, A., (2018). Development of lamivudine containing multiple emulsions stabilized by gum odina. *Future J. Pharm. Sci., 4*, 71–79.

Juqun, X., Ling, Z., & Hua, D., (2012). Drug-loaded chondroitin sulfate-based nanogels: Preparation and characterization. *Colloids Surf. B: Biointerf., 100*, 107–115.

Katia, M., Enrica, M. T., Miriam, S., Paola, P., Bice, C., Tiziana, M., & Franca, P., (2009). *In vitro* evaluation of chondroitin sulfate-chitosan microspheres as a carrier for the delivery of proteins. *J. Microencapsul., 26*(6), 535–543.

Katiuscia, V. J., Graziella, A. J., Ricardo, B. A., & Alexandre, L. P., (2015). Physico-chemical characterization and cytotoxicity evaluation of curcumin loaded in chitosan/chondroitin sulfate nanoparticles. *Material. Sci. Eng. C., 56*, 294–304.

Khalid, I. , Ahmad, M., Minhas, M.U., & Barkat, K., (2018). Synthesis and evaluation of chondroitin sulfate based hydrogels of loxoprofen with adjustable properties as controlled release carriers. *Carbohydr. Polym., 181*, 1169–1179.

Kwok, J. C. F., Warren, P., & Fawcett, J. W., (2012). Chondroitin sulfate: A key molecule in the brain matrix. *Int. J. Biochem. Cell. Biol., 44*, 582–586.

Miller, K. L., & Clegg, D. O., (2011). Glucosamine and chondroitin sulfate. *Rheum. Dis. Clin. N. Am., 37*, 103–118.

Ming, F., Ye, M., Huaping, T., Yang, J., Siyue, Z., Shuxuan, G., Meng, Z., Hao, H., Zhonghua, L., Yong, C., & Xiaohong, H., (2017). Covalent and injectable Chitosan-chondroitin sulfate hydrogels embedded with chitosan microspheres for drug delivery and tissue engineering. *Material. Sci. Eng. C., 71*, 67–74.

Miyata, S., & Kitagawa, H., (2017). Formation and remodeling of the brain extracellular matrix in neural plasticity: Roles of chondroitin sulfate and hyaluronan. *Biochim. Biophys. Acta., 1861*(10), 2420–2434.

Mohtashamiana, S., Boddohia, S., & Hosseinkhani, S., (2018). Preparation and optimization of self-assembled chondroitin sulfate-nisin nanogel based on quality by design concept. *Int. J. Biol. Macromol., 107B*, 2730–2739.

Nayak, A. K., & Pal, D., (2012). Natural polysaccharides for drug delivery in tissue engineering. *Everyman's Sci., XLVI*, 347–352.

Nayak, A. K., & Pal, D., (2015). Chitosan-based interpenetrating polymeric network systems for sustained drug release. In: Tiwari, A., Patra, H. K., & Choi, J. W., (eds.), *Advanced Theranostics Materials* (pp. 183–208). WILEY-Scrivener, USA.

Nayak, A. K., & Pal, D., (2016a). Sterculia gum-based hydrogels for drug delivery applications. In: Kalia, S., (ed.), *Polymeric Hydrogels as Smart Biomaterials* (pp. 105–151). Springer series on polymer and composite materials, Springer International Publishing, Switzerland.

Nayak, A. K., & Pal, D., (2016b). Plant-derived polymers: Ionically gelled sustained drug release systems. In: Mishra, M., (ed.), *Encyclopedia of Biomedical Polymers and Polymeric Biomaterials* (Vol. VIII, pp. 6002–6017). Taylor & Francis Group, New York, NY 10017, U.S.A.

Nayak, A. K., & Pal, D., (2017a). Natural starches-blended ionotropically-gelled microparticles/beads for sustained drug release. In: Thakur, V. K., Thakur, M. K., & Kessler, M. R., (eds.), *Handbook of Composites from Renewable Materials* (Vol. 8, pp. 527–560). Nanocomposites: Advanced applications, WILEY-Scrivener, USA.

Nayak, A. K., & Pal, D., (2017b). Tamarind seed polysaccharide: An emerging excipient for pharmaceutical use. *Indian J. Pharm. Educ. Res., 51*, S136–S146.

Nayak, A. K., & Pal, D., (2018). Functionalization of tamarind gum for drug delivery. In: Thakur, V. K., & Thakur, M. K., (eds.), *Functional Biopolymers* (pp. 35–56). Springer International Publishing, Switzerland.

Nayak, A. K., (2016). Tamarind seed polysaccharide-based multiple-unit systems for sustained drug release. In: Kalia, S., & Averous, L., (eds.), *Biodegradable and Bio-Based Polymers: Environmental and Biomedical Applications* (pp. 471–494). WILEY-Scrivener, USA.

Nayak, A. K., Ara, T. J., Hasnain, M. S., & Hoda, N., (2018a). Okra gum-alginate composites for controlled releasing drug delivery. In: Inamuddin, A. A. M., & Mohammad, A., (eds.), *Applications of Nanocomposite Materials in Drug Delivery* (pp. 761–785). Elsevier Inc.

Nayak, A. K., Bera, H., Hasnain, M. S., & Pal, D., (2018c). Graft-copolymerization of plant polysaccharides. In: Thakur, V. K., (ed.), *Biopolymer Grafting, Synthesis and Properties* (pp. 1–62). Elsevier Inc.

Nayak, A. K., Hasnain, M. S., & Pal, D., (2018b). Gelled microparticles/beads of sterculia gum and tamarind gum for sustained drug release. In: Thakur, V. K., & Thakur, M. K., (eds.), *Handbook of Springer on Polymeric Gel* (pp. 361–414). Springer International Publishing, Switzerland.

Nayak, A. K., Pal, D., & Santra, K., (2015). Screening of polysaccharides from tamarind, fenugreek and jackfruit seeds as pharmaceutical excipients. *Int. J. Biol. Macromol., 79*, 756–760.

Nayak, A. K., Pal, D., & Santra, K., (2016). Swelling and drug release behavior of metformin HCl-loaded tamarind seed polysaccharide-alginate beads. *Int. J. Biol. Macromol., 82*, 1023–1027.

Pal, D., & Nayak, A. K., (2015a). Alginates, blends and microspheres: Controlled drug delivery. In: Mishra, M., (ed.), *Encyclopedia of Biomedical Polymers and Polymeric Biomaterials* (Vol. I, pp. 89–98). Taylor & Francis Group, New York, NY 10017, U.S.A.

Pal, D., & Nayak, A. K., (2015b). Interpenetrating polymer networks (IPNs): Natural polymeric blends for drug delivery. In: Mishra, M., (ed.), *Encyclopedia of Biomedical Polymers and Polymeric Biomaterials* (Vol. VI, pp. 4120–4130). Taylor & Francis Group, New York, NY 10017, U.S.A.

Pal, D., & Nayak, A. K., (2017). Plant polysaccharides-blended ionotropically-gelled alginate multiple-unit systems for sustained drug release. In: Thakur, V. K., Thakur, M. K., & Kessler, M. R., (eds.), *Handbook of Composites from Renewable Materials* (Vol. 6, pp. 399–400). Polymeric composites, WILEY-Scrivener, USA.

Rajan, M., & Raj, V., (2013). Formation and characterization of chitosan-polylactic acid-polyethylene glycol-gelatin nanoparticles: A novel biosystem for controlled drug delivery. *Carbohydr. Polym., 98*, 951–958.

Sherman, A. L., Ojeda-Correal, G., & Mena, J., (2012). Use of glucosamine and chondroitin in persons with osteoarthritis. *Phys. Med. Rehab., 4*, 110–116.

Silver, D. J., & Silver, J., (2014). Contributions of chondroitin sulfate proteoglycans to neurodevelopment, injury, and cancer. *Curr. Opin. Neurobiol., 27*, 171–178.

Sleep, D., (2014). Albumin and its application in drug delivery. *Expert Opin. Drug Deliv., 12*, 793–812.

Umerska, A., Corrigan, O. I., & Tajber, L., (2017). Design of chondroitin sulfate-based polyelectrolyte nanoplexes: Formation of nanocarriers with chitosan and a case study of salmon calcitonin. *Carbohydr. Polym., 156*, 276–284.

Xu, M., Wei, L., Xiao, Y., Bi, H., Yang, H., & Du, Y., (2017). Physicochemical and functional properties of gelatin extracted from Yak skin. *Int. J. Biol. Macromol., 95*, 1246–1253.

CHAPTER 6

Biodegradability and Biocompatibility of Natural Polymers

ABUL K. MALLIK, MD SHAHRUZZAMAN, MD SAZEDUL ISLAM, PAPIA HAQUE, and MOHAMMED MIZANUR RAHMAN

Department of Applied Chemistry and Chemical Engineering, Faculty of Engineering and Technology, University of Dhaka, Dhaka 1000, Bangladesh

ABSTRACT

Polymers are one of the most important materials used in pharmaceutical industries. Various types of polymers (especially natural polymer-derivatives) can be used for the preparation of cosmetics and drugs, and their applicability could be increased by changing the structure and physical properties. It is also reported to the development of polymers for targeted drug delivery system. Naturally derived polymers have more advantages compared to the synthetic one, such as biodegradability, biocompatibility, and biological activity, because tissues of living organisms have these properties. To be a biodegradable polymer, the most important requirement is to be applied in medical purpose is its compatibility regarding physical and chemical properties as well as their behavior when they come into contact with the body. One of the requirement of a material to be used as biomaterials, it should have biocompatibility. Generally, biocompatibility is the connection between a material and the organism where none of them yields adverse effects. A biocompatible material should pass various studies extending from in vitro assays to clinical trials in various fields (pharmaceutics, toxicology, chemistry, biology, etc.). This chapter will be focused on biodegradability and biocompatibility of various natural polymers.

6.1 INTRODUCTION

Generally, a polymer is defined as a big molecule composed of many repeating units linked by covalent chemical bonds. It can be synthetic or natural-origin however; natural-origin polymers are readily available, economical, and non-toxic. Moreover, they have other advantages compared to synthetic ones, such as biodegradability, biocompatibility, and biological performance, because of their common existence in the tissues of living organisms (Satturwar et al., 2003). Therefore, they are very attractive in pharmaceutical applications. The most important condition to be a biodegradable polymer to be applied in medical purpose is its compatibility regarding physical and chemical characteristics as well as their behavior when they come into interaction with the body (Silva et al., 2004). A biomaterial should have the characteristics of biocompatibility which implies if the material use in any organism they will be compatible without creating any problem. Natural-based polymers can be divided into two groups; namely polysaccharides and protein-origin.

One of the most important classes of biopolymers are polysaccharides contained simple sugar monomers (Nishinari and Takahashi, 2003). Polysaccharides can form linear and branched type structure where monomers or monosaccharides are connected together by O-glycosidic bonds. Sources of these polysaccharides could be animal, plant, or microbial (Cascone et al., 2001). Polysaccharides have many advantages, for example, they have hemocompatibility, which may be due to their chemical similarities with heparin. These biopolymers are also low in cost, interacts with living cells, and non-toxic (Cascone et al., 2001; Venugopal and Ramakrishna, 2005). On the other hand, proteins are biopolymers containing 20 different amino acids, and they are linked by amide/peptide bonds (Mallik et al., 2018). Therefore, amino acids are the building blocks for proteins, and their properties also varied with the existing amino acids. Protein-derived biopolymers have the benefits of mimicking numerous properties of extracellular matrix (ECM). Consequently, they have the capacity to direct the growth, migration, and arrangement of cells throughout tissue regeneration as well as wound healing. Encapsulated and transplanted cells also can be stabilized by them (Malafaya et al., 2007). These aforementioned biomaterials from natural origin have potential applications in biomedical and pharmaceutical industries.

Biomaterial can be explained as a substance of natural or synthetic origin, which can interrelate with the biological systems and direct medical treatment (Williams, 2009). They are considered as nonviable materials which can be a part of the body and work as a natural tissues and/or organ (Park and Park, 1996). Therefore, in many medical applications biomaterials have been used, such as orthopedic devices (Chasin et al., 1988), controlled drug delivery (Cohen et al., 1991), vascular grafts and cardiac pacemakers (Piecuch and Fedorka, 1983). When introduced *in vivo*, the biomaterials should have the properties not to produce any carcinogenic, immunologic, systemic, cytotoxic mutagenic, or teratogenic reactions (Ratner et al., 1975).

On the other hand, usually active drugs are formulated into a suitable dosage form with the aid of some excipients and the excipients have many functions like lubricating, gelling, binding, suspending, bulking agent, flavoring, sweetening, etc. Excipients are very important in the preparation of medicines because they help to maintain the safety, efficacy, and constancy of active pharmaceutical constituents. They are the highest constituents of any pharmaceutical formulation originated from natural or synthetic sources. Applications of natural polymers as raw materials in the pharmaceutical industry play a vital role due to their biodegradability and nontoxicity. They also cost-effective and comparative abundance to synthetic polymers (Malafaya et al., 2007; Malviya et al., 2011). Moreover, natural resources are renewable and sustainable, which can offer a constant supply of raw material (Perepelkin, 2005). Polysaccharides and their derivatives are extensively applied in pharmaceutical preparations, and in various cases, their existence are very important help to release the drug in an appropriate rate. Pharmaceutical industries have been using naturally occurring polymers as excipients in various dosage forms (solid, liquid, semisolid, etc.) for their pharmaceutical formulations.

In this chapter, the various natural-derived materials that have biodegradability and biocompatibility properties will be discussed in details. The biomedical and pharmaceutical applications of natural polymers will also be overviewed.

6.2 BIODEGRADABLE NATURAL POLYMERS

Natural polymers are very well-known and efficient materials due to their abundance, biocompatibility, biodegradability, and environmental concerns. The demand of natural polymeric materials is increasing over the last two decades because of their biocompatibility and non-toxicity compare to most of the other synthetic polymers. This has led natural polymeric materials suitable to prepare various kinds of biomaterials that have been widely used in different fields. Natural polymers are usually made from plant materials that can be grown-up indefinitely. Large structures of natural polymer can be formed when the monomeric units are covalently bonded to each other. Biodegradable natural polymers are mainly of three kinds: polynucleotides (RNA and DNA), polypeptides, and polysaccharides (Meyers et al., 2008; Kumar et al., 2007). Biodegradable natural polymers are mainly based on renewable sources such as cellulose, starch, hyaluronan, alginate, chitosan, etc. These biodegradable polymers are widely applied in various fields such as pharmaceutical, agriculture, wastewater treatment, cosmetics, removal of heavy metals, and so on. Compared to synthetic polymers, natural polymers exhibit better biode-gradability and biocompatibility because of enzymatic biodegradation. Lysozyme, one of the most important human enzymes is responsible for such kind of enzyme-controlled biodegradation that produces biocompat-ible byproducts. In the following sections, details about biodegradability and biocompatibility of natural polymers (polysaccharide- and protein-based) and their applications in pharmaceutical sectors will be discussed.

6.2.1 POLYSACCHARIDES

The ever-increasing interest to utilize polysaccharides as biodegradable materials has gained increased attention because polysaccharide-based materials exhibit various inherent properties that can be used in several applications. In this context, they have opened new roads in the phar-maceutical domain, due to their unique benefits, like non-toxicity, abun-dance, biodegradability, biocompatibility, and biological functions. Some examples of polysaccharides are cellulose, alginate, chitosan, starch, hyaluronan, gum, etc. Various kinds of polysaccharides and their pharma-ceutical applications are discussed in the following subsections.

6.2.1.1 CELLULOSE AND ITS DERIVATIVES

At present, cellulose is the most abundant renewable and biodegradable polymer on earth because the yearly production of cellulose is estimated of 1.5×10^{12} tons. The cell walls of lignocellulosic plants are considered as the main component of cellulose. From the view of chemistry, the polymer chain of cellulose consists of unbranched β-(1→4) glycosidic linked D-glucose units with hydroxyl groups arranged in an equatorial nature (Figure 6.1). Nowadays, cellulose and its derivatives-based materials have gained a much more popularity due to their environmentally friendly nature compared to conventional materials. Moreover, the excellent inherent properties of cellulose and its derivatives-based material make them very attractive in many applications (Klemm et al., 2005) such as agriculture, textile industry, pharmaceutical industry, and so on.

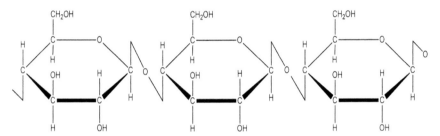

FIGURE 6.1 Structure of cellulose.

Cellulose has gained much more popularity than any other biodegradable polymer because the β-(1→4) glycosidic linkage of cellulose is attacked by various types of microorganisms. It is obvious that pure cellulose decomposes readily (Chandra and Rustgi, 1998) as cellulose and lignin exists together in the structure. Biocompatibility of cellulose and its derivatives was examined by Miyamoto et al., in two *in vivo* tests, one of them is for absorbance by living tissue, and the other is for foreign body reaction.

Excellent biocompatibility of cellulose and cellulose derivatives determine their wide-ranging applications in pharmaceutical compounded and industrialized products. Cellulose esters and cellulose ethers are the two main groups of cellulose derivatives having different physicochemical and mechanical properties. Thus, water-insoluble cellulose esters compounds having good film-forming characteristics find a variety of applications such as material-coatings and controlled-release systems, hydrophobic matrices, and semipermeable membranes in pharmaceuticals, agriculture, and cosmetics. Additionally, cellulose esters such as cellulose acetate, cellulose acetate phthalate, cellulose acetate butyrate, cellulose acetate trimelitate, and hydroxypropylmethylcellulose (HPMC) phthalate are used extensively as binders, fillers, and laminate layers in composites and laminates. Also, they are used as outstanding material for photographic films, and as membrane-forming materials applicable for gas separation, water purification, food, and beverage processing, medicinal, and bioscientific fields (Malhotra et al., 2015; Kusumocahyo et al., 2005). On the contrary, cellulose ethers that present a good solubility, high chemical resistance, and non-toxicity are used in drilling technology and building materials, as stabilizers in food, pharmaceutical, and cosmetics formulations as the main component. Moreover, Matrix tablets design is another important application of cellulose ethers, where cellulose ethers are used as excipients (Colombo et al., 1999). The most commonly used cellulose ethers are: methylcellulose (MC), ethyl cellulose (EC), hydroxyethyl cellulose (HEC), carboxymethyl cellulose (CMC), sodium carboxymethyl cellulose (NaCMC), hydroxypropyl cellulose (HPC) and HPMC.

There are some conditions under which chemically modified celluloses show biodegradability. In the esterification procedure, the glucopyranosyl hydroxyl groups of varying degrees are replaced by hydrophobic ester groups. Amass et al., (1998) reported that the degrees of substitution and C-2 hydroxyl substitution patterns are important standards to predict the biodegradation patterns of these polymers.

6.2.1.2   CHITIN/CHITOSAN

Chitin, a fibrous substance consisting of polysaccharides, is found in the shells of arthropods such as crustaceans (e.g., lobsters, crabs, and shrimps). And chitosan is a unique bio-based polymer that is the derivative of chitin and forms the exoskeleton of arthropod. Chitin can be converted to chitosan by partial deacetylation using chemical method or enzymatic hydrolysis (Jayakumar et al., 2010). The chemical composition of chitin and chitosan obtained from shrimp head and crab shell are listed in Table 6.1.

Chitosan is the second most copious natural polysaccharide next to cellulose that consists of $\beta(1–4)$ linked D-glucosamine with randomly located N-acetylglucosamine groups (Figure 6.2) (Nair and Laurencin, 2007). Chitosan is soluble in water; however, chitin is insoluble in its

Biodegradability and Biocompatibility of Natural Polymers 139

native form. The biocompatibility, biodegradability, and non-toxicity of chitosan make them strong candidate to use in tissue engineering, biosensors, oral administration in humans, drug delivery and wound healing, wastewater treatment, removal of heavy metals, food additives and so on (Duceppe and Tabrizian, 2010; Sezer and Cevher, 2012).

TABLE 6.1 Chemical Composition of Chitin and Chitosan Obtained from Shrimp Head and Crab Shell[*]

Chemical Composition	Shrimp Head		Crab Shell	
	Chitin (%)	Chitosan (%)	Chitin (%)	Chitosan (%)
Moisture content	18.90	11.90	16.70	7.90
Crude protein	< 0.10	< 0.10	< 0.10	< 0.10
Crude lipid	1.90	2.20	2.80	2.50
Crude ash	4.90	3.60	4.90	5.50
Degree of deacetylation	10.26	0.20	5.40	0.10
Viscosity	1.18	1.42	1.04	1.35
Yield	23.3	12	25	65
Particle size	< 1 mm	< 1 mm	< 1 mm	< 1 mm
Color	Pink	Pink	Ivory	Ivory

[*]Food and Fertilizer Technology Centre (FFTC) Document Database

The biocompatibility of chitosan scaffolds were examined by Vande Vord et al., (2002) using a mammalian implantation model. They reported the early migration of neutrophils into the implantation area with increase the time of implantation. In their study, they observed the immune reactions by lymphocyte proliferation assays and antibody binding responses by using ELISA techniques. In addition, they also observed the typical healing response by the formation of normal granulation tissue associated with accelerated angiogenesis. It was concluded from the results of their study that chitosan shows higher degree of biocompatibility.

The *in vitro* degradation of chitosan occurs due to enzymes named chitosanase, lysozyme, and papain. Lysozyme primarily degrades chitosan *in vivo*; however, the degradation finally takes place through the hydrolysis of the acetylated residues. Shi et al., and his co-workers (Shi et al., 2006) reported that the chitosan degradation rate inversely depends on two things, one is the degree of acetylation and another one is the crystallinity of the polymer. The lowest degradation rates were exhibited by the highly deacetylated form and may last several months *in vivo*. Besides, the solubility and degradation rate was considerably affected by the chemical modification of chitosan. In addition, chitosan molecule has amino and hydroxyl groups which can be modified chemically providing materials that have a variety of physical and mechanical properties. For example, suitable biological and physical properties of porous chitosan matrix make it an effective material for bone regeneration.

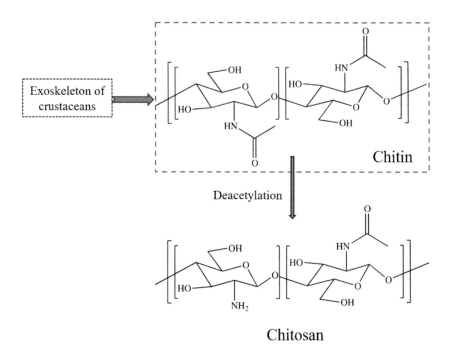

FIGURE 6.2 (See color insert.) Deacetylation of chitin to chitosan.

6.2.1.3 STARCH

Starch is one of the most abundant storage polysaccharide that is found in the chloroplasts of plant cells as insoluble granules. From a chemical point of view, starch is composed of linear α-amylose polymer (20–30%) that contains several thousands of glucose residues linked by α(1→4) bonds and branched amylopectin (70–80%) that consists mainly of α(1→4)-linked glucose residues with α(1→6) branch points at every 24–30 glucose residues in average (Figure 6.3). Amylopectin molecules are the largest molecules in nature because the molecules contain up to 10^6 glucose residues (Voet et al., 1999).

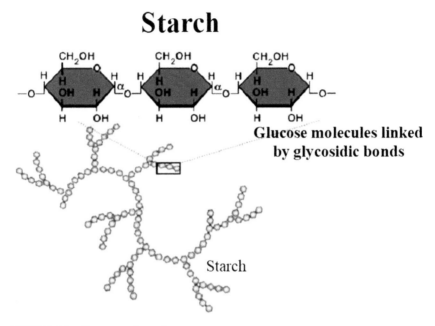

FIGURE 6.3 Structure of starch.

Because of its inherent biodegradability, biocompatibility, and non-toxicity, starch has gained much more consideration for the development of renewable materials combining with various types of natural as well as synthetic sources. Baran et al., (2004) reported that starch-based polymers show better performance when used in the packaging, drug delivery systems, and tissue engineering scaffolds. However, poor processability, dimensional stability, low shear resistance, retrogradation tendency, and high moisture sensitivity make starch restricted to apply in various sectors. For that reason, starch is associated with either biopolymers or synthetic polymers to retain the biocompatibility and biodegradability of the final product (Chatakanonda et al., 2000).

Reis and his co-workers proposed pharmaceutical and biomedical applications of starch-based materials (blends of starch with different synthetic polymers, such as ethylene vinyl alcohol, polylactic acid, cellulose acetate, and polycaprolactone). They reported that these types of materials have been extensively used for drug delivery applications, including cancer therapy, nasal administration, porous scaffolds for bone tissue engineering, and so on. These materials have been shown to be biocompatible in vitro (Marques et al., 2002), and to possess a good in vivo performance (Salgado et al., 2007). The aim of their research was to evaluate starch-based polymers and composites by screening their cytotoxicity to proof them as potential biomaterials. For the assessment of biocompatibility, cytotoxicity, and cell adhesion tests were done for two different blends of corn-starch, starch ethylene vinyl alcohol (SEVA-C) and starch cellulose acetate (SCA) and their respective composites with hydroxyapatite (HA). From the result of their study, it was concluded that both types of starch-based polymers exhibit a cytocompatibility and can be used as biomaterials. In another study, A.J. Salgado et al., evaluate the biocompatibility of three starch-based scaffolds by implanting them in rats to confirm *in vivo* endosseous response.

6.2.1.4   ALGINATE

Alginate is a broadly distributed linear polysaccharide and copolymers with homopolymeric blocks of (1→4) linked β-D-mannuronic acid (M) and α-l-guluronic acid (G) monomers of different composition and sequence (Figure 6.4). They are abundant in nature and are found in the cell wall of brown algae. Three brown algae species, including *Laminaria hyperborean*, *Ascophyllum nodosum*, and *Macrocystis pyrifera* are used to extract commercial alginates. Alginates have been extensively used as a effective biomaterial for many years due to its biocompatibility, low toxicity, relatively low cost, and mild gelation. Alginates are many used in biomedical sectors such as in tissue engineering, wound healing, cartilage, and bone regeneration, protein delivery, *in vitro* cell culture, antibacterial films, drug delivery (Draget and Taylor, 2011), etc. However, the lacking of good mechanical and thermal properties of alginates limits their use in various applications. To minimize the limitation, alginate is often associated with natural as well as synthetic polymers to prepare potential materials with improved mechanical and thermal properties.

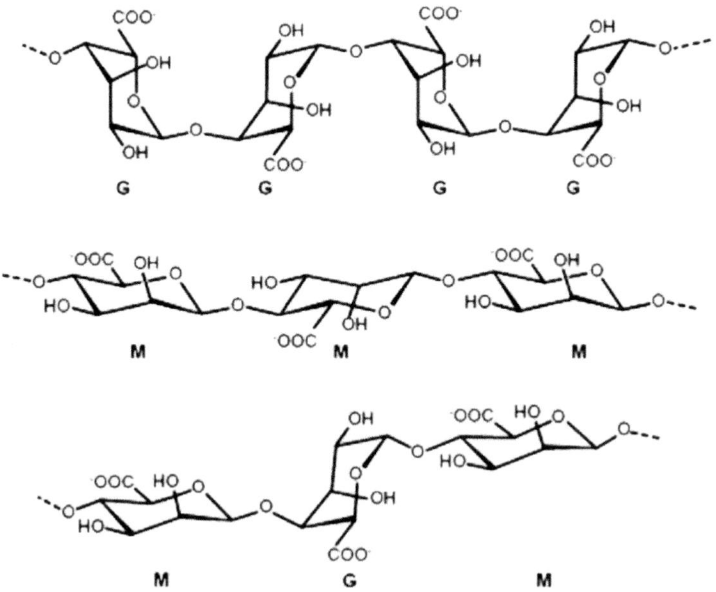

FIGURE 6.4   Chemical structures of G-block, M-block, and alternating block in alginate. Reprinted with permission from Elsevier (Lee and Mooney, 2012).

Biocompatibility of biomaterials is a vital matter for the long-term function on multiple therapeutic systems. Many researchers examined biocompatibility of alginates and alginate-based materials. G. Orive et al., presented in vitro techniques to assess the biocompatibility of alginates and alginate-based microcapsules with different compositions and purities (Orive et al., 2005). In another study, Ulrike Rottensteiner et al., evaluated *in vitro* and *in vivo* biocompatibility of alginate-based material used as a scaffold in bone tissue engineering (Rottensteiner et al., 2014). There *in-vitro* biocompatibility evaluation indicated that alginate dialdehyde-gelatin (ADA-GEL) hydrogel films are good candidate for bone tissue engineering as it show good cell adherence and proliferation of bone marrow-derived mesenchymal stem cells (MSC).

6.2.1.5   HYALURONAN

Hyaluronan, also known as hyaluronic acid (HA), is an anionic, non-sulfated glycosaminoglycan and a major macromolecular component distributed widely throughout connective, epithelial, and neural tissues (Liao et al., 2005). From the chemical point of view, hyaluronan is a linear polysaccharide consisting alternating disaccharide units of α-1,4-D- glucuronic acid and β-1,3-N-acetyl-D-glucosamine that is linked by β(1→3) bonds (Laurent et al., 1995). Umbilical cord, rooster comb, synovial fluid, or vitreous humor is the main sources to extract commercial hyaluronan. In addition, hyaluronan can easily be produced in large scales from strains of bacteria (Streptococci) by using microbial fermentation process (Laurent et al., 1995). A number of physiological and macromolecular functions are covered by hyaluronan networks such as tissue and matrix water regulation, structural, and space-filling properties, lubrication, etc. Hyaluronan has been extensively used in biomedical arena, including drug delivery, gene delivery, implantable delivery devices, and so on (Laurent et al., 1995).

Biocompatibility and *in vitro* cytotoxicity test of modified hyaluronan scaffold was performed by Lubos Danisovic and his co-workers. For their study, they used direct contact assay as well as MTT test where human adipose tissue-derived stem cells were used as biological model (Danisovic et al., 2013). It was summarized that hyaluronan modified scaffold is non-toxic and biocompatible and should be used as carrier of various types of cells. In another study, Sungchul Choi et al., evaluated *in vitro* biocompatibility by measurement of cell counting and assay of cell live and dead (Choi et al., 2014). The cell viability and cell proliferation measurement study concluded that the purified hyaluronan solution showed excellent cell compatibility with zero cell damages.

6.2.1.6   CHONDROITIN SULFATE

Chondroitin sulfate is an unbranched polysaccharide that is normally found in cartilage around joints in the body, synovial fluid, bone, and heart valves. When these polysaccharides are covalently linked to a protein core, proteoglycans is formed (Pieper et al., 1999). Chondroitin sulfate (CS) is a sulfated glycosaminoglycan consists of two alternating monosaccharides, D-glucuronic acid and N-acetyl galactosamine at either 4- or 6-positions. It is commercially available in USA and Europe market as a nutritional supplement. CS together with glucosamine is used as dietary supplement for the treatment of osteoarthritis. It is also used as skin substitute for

Biodegradability and Biocompatibility of Natural Polymers 145

treating burns (Lee et al., 2005). However, soluble nature of CS in water restricts its biomedical application as like drug delivery.

The *in vitro* studies reported by Bassleer et al., showed that matrix component production by human chondrocytes is increased due to the presence of CS (Bassleer et al., 1998). Moreover, to facilitate the design of delivery systems, negatively charged CS interacts with positively charged molecules such as polymers. Furthermore, Jeong Park et al., reported the preparation of CS–chitosan sponges, which was used for bone regeneration (Park et al., 2000).

6.2.1.7   GUMS

Gums are a group of polysaccharides that are produced at the surface of a plant upon the introduction of counterions (Chandra and Rustgi, 1998). Most of the gums have two important characteristics: 1) gums are soluble in water and 2) gums have high viscosity. Food and pharmaceutical industries are the two major areas where gums are used because of their emulsifying, stabilizing, thickening, and gel-forming properties. Larch gum, gum Arabic, gum karaya, guar gum, and gum tragacanth are the major types of gums now commercially available in the market (Stephen et al., 1990).

Arabinogalactan, a type of D-galactan is obtained from softwoods called larch (larch gum). Larch gum, extracted from the Larix tree is composed of D-galactose units linked by $\beta(1 \rightarrow 3)$ linkage containing a side chain at position C6. Due to its biocompatibility, biodegradability, high water solubility, and the ease of chemical modification, larch gum is an eye-catching polymer for the synthesis of scaffolds for tissue engineering application.

Gum Arabic, also known as acacia gum, is obtained from various species of the acacia trees. It is a natural gum consisting of a variable mixture of arabinogalactan oligosaccharides, polysaccharides, and glycoproteins. $\beta(1 \rightarrow 3)$ and $\beta(1 \rightarrow 6)$-linked D-galactose units along with $\beta(1 \rightarrow 6)$-linked D-glucopyranosyl uronic acid units are the foundations of the main chain of this polysaccharide. Gum Arabic shows emulsification, encapsulation, and film-forming properties due to its high water solubility (up to 50% w/v) and relatively low viscosity characteristics (Izydorczyk et al., 2005). Gum Arabic is biocompatible in nature, and for that reason, this polymer has been tested by several groups (Wilson Jr. et al., 2008) for the surface modification of magnetic nanoparticles for different applications.

Gum karaya, also known as sterculia gum, is a vegetable gum taken from the tree *Sterculia urens*. People use it to make medicine. Gum karaya is applied as a laxative to relieve constipation, in cosmetics and denture adhesives, and as a thickener and emulsifier in foods. It is used as a bulk-forming laxative to relieve constipation. Gum karaya stimulates the digestive tract by swelling the intestine, which results to push stool through.

Guar gum, also known as guaran, is a hydrophilic and nonionic galactomannan polysaccharide, derived from guar kernels (*Cyamopsis tetragonolobus*) that come from *Leguminosae* family. In a chemical point of view, the backbone of guar gum is a linear chain of $(1{\rightarrow}4)$-linked β-D-mannopyranosyl units along with $(1{\rightarrow}6)$-linked α-D-galactopyranosyl residues as side chains. Due to its biodegradability, non-toxicity, low-cost, availability, and stabilizing properties, guar gum can be extensively used in pharmaceutical formulations, textile, petroleum, paper, cosmetic, food, toiletries (Prabhanjan et al., 1989), etc.

6.2.1.8 OTHER POLYSACCHARIDE-BASED POLYMERS

Other polysaccharides such as dextran, agar, carrageenan, etc. have been studied for pharmaceutical and other applications. Dextran, discovered by Louis Pasteur, is a complex branched polysaccharide made of many D-glucose molecules (Naessens et al., 2005). Dextran is a biodegradable and biocompatible polymer consists of $\alpha(1{\rightarrow}6)$-linked D-glucose residues with some degree of branching via $\alpha(1{\rightarrow}3)$ linkages. Biodegradability and biocompatibility of dextran make it suitable for use in a wide range of applications, such as treatment of hypovolemia, as bone healing promoter, drug delivery and also for dermal and subcutaneous augmentation (Cascone et al., 2001).

Agar, the linear polysaccharide, is a jelly-like substance composed of two components named agarose and agaropectin. It is a unique natural polymer that increasingly preferred over synthetic materials, in addition, to be considered alternative sources of raw materials for pharmaceutical applications. Due to its outstanding intrinsic properties such as gel strength and microbial flora absence, agar is used as gelling, stabilizer, and thickening agent in the pharmaceutical industry.

Carrageenan, a family of sulfated polysaccharide, can be manufactured from red marine algae such as *Kappaphycus alvarezii*. At room temperature, carrageenan can form stiff and thermo-reversible gels in the presence of gel-promoting salts (Mangione et al., 2005). It is a natural polymer consists of repeating units of (1,3)-D-galactopyranose and (1,4)-3,6-anhydro-α-D-galactopyranose with sulfate groups in a certain amount and position (Campo et al., 2009). Because of its non-toxicity, biodegradability, biocompatibility, and gel-forming ability, carrageenan is favorable biopolymer to use as a thickener in food and nonfood industries.

6.2.2 PROTEIN-ORIGIN POLYMERS

Protein-based polymers are usually composed of repeated units of amino acids and have emerged as an attractive class of natural polymers because of their desirable mechanical, chemical, and biological properties like biocompatibility, biodegradation (Meyer and Chilkoti, 2002). Moreover, the molecular weight and sequence of amino acid of these polymers can be controlled precisely, which ultimately determines the pharmacokinetics, biological activity, and biodegradation of the polymer. As a result, these polymers become popular in fabrication of biomaterials for a wide range of applications like drug delivery system, tissue engineering, etc.

6.2.2.1 COLLAGEN

Collagen is the primary structural material in soft and hard connective tissues like skin, bone, cartilage, tendon, etc. and most abundant protein (about 20–30% of total body proteins) in mammals (Harkness, 1966). To date, 29 different types of collagen have been identified in the human body, with the most common being Type I–IV (Gorgieva and Kokol, 2011). Collagen molecules are comprised of three polypeptide chains with unique amino acid sequence, twined around one another as in a three-stranded rope which are held together primarily by hydrogen bonds between adjacent -CO and -NH groups, but also by covalent bonds (Harkness, 1966).

Biodegradation of collagen follows the pathway of enzymatic degradation within the body by enzymes, such as collagenases and metalloproteinases, forming collagen fragments which are further degraded by gelatinases and nonspecific proteases (Aimes and Quigley, 1995). The peptide bonds of collagen in a triple helix are occluded from enzyme's active sites, but the single-stranded regions are susceptible to cleavage by matrix metalloproteinases (MMPs). As a result of degradation of collagen, chemotaxis of human fibroblasts is produced (Postlethwaite et al., 1978) which are believed to promote restoration of tissue structure and functionality (Yannas et al., 1982).

Physico-chemical, mechanical, and biological properties of collagen largely depends on its enzymatic degradation and to be used in different biomedical application. Low biomechanical stiffness and high rate of enzymatic degradation of natural collagen causes rapid biodegradation, which is a disadvantage of collagen-based biomaterials. The degradation rate of collagen can be controlled by enzymatic pre-treatment or cross-linking with various cross-linkers (Nair and Laurencin, 2007). Different cross-linkers like difunctional or multifunctional aldehydes, polyepoxy compounds, carbodiimides, hexamethylene-diisocyanate, and succinimidyl ester polyethylene glycol, etc. can serve the purpose.

Song et al., studied the biocompatibility and biodegradation behavior of collagen extracted from Jellyfish (Song et al., 2006). To control the enzymatic degradation kinetics, jellyfish collagen scaffolds were crosslinked with 1-ethyl-(3-3 dimethylaminopropyl) carbodiimide hydrochloride/N-hydroxysuccinimide (EDC/NHS). The enzymatic degradation behavior of cross-linked and non-cross-linked collagen scaffolds was investigated by observing the residual mass percent of the matrix as a function of time. It was found that the degradation of non-cross-linked collagen scaffolds was rapid while cross-linking with EDC/NHS led to a significant improvement in biostability. It is possible to evaluate the biocompatibility of implanted biomaterials into our body by measuring the duration and intensity of the immune responses against implanted biomaterials. When foreign materials or pathogens enter into our immune system, it starts to secrete a large amount of antibodies and cytokines. Song et al., measured the level of proinflammatory cytokine secretions and antibody secretions, and changes in number of immune cells were monitored after in vivo implantation to evaluate the biocompatibility of jellyfish derived collagen. The results revealed low immune response of collagen scaffolds, which indicated its high biocompatibility (Song et al., 2006).

The in vivo biocompatibility and degradation behavior of thin collagen-based cell carrier (CCC) rat animal model was evaluated by Rahmanian-Schwarz et al., The results revealed no evidence of encapsulation, hypertrophic scar formation or long-term foreign body reaction and

Biodegradability and Biocompatibility of Natural Polymers 149

inflammation introduced by the implanted CCC and thus confirmed its high biocompatibility (Rahmanian-Schwarz et al., 2014). The implanted CCC also showed low irritability, complete resorption, and replacement by autologous tissue, which encouraged its application in surgical applications and regenerative medicine. The implanted CCC was completely degraded after 42 days of subcutaneous implantation.

6.2.2.2 GELATIN

Gelatin is a water-soluble protein obtained by thermal denaturation or physical and chemical degradation of collagen animal skins, bones, and tendons which involves the breaking of the triple-helix structure resulting in helix-to-coil transition and conversion into soluble gelatin. Molecular weight of gelatin varies due to the differences in collagen sources and the conditions of extraction of the gelatin, which ultimately controls its structure, physical properties, and chemical heterogeneity. Gelatin is found to contain about 18 amino acids linked together in a partially ordered fashion and out of these glycine or alanine accounts for about one-third to half of the total amino acid residues.

Products from biodegradation of gelatin, i.e., gelatin-derived hydrolysates and peptides are generally obtained by enzymatic proteolysis which can be achieved by a number of proteases including trypsin, chymotrypsin, pepsin, alcalase, properase E, pronase, collagenase, bromelain, and papain, etc. (Kim et al., 2001). Like collagen, it is also useful to crosslink gelatin to increase its strength and enzyme resistance and to maintain their stability during implantation, especially for long term application. A number of crosslinkers are available to stabilize gelatin like genipin, glutaraldehyde, diisocyanates, carbodiimides, polyepoxy compounds, acyl azide, etc.

Biocompatibility of a gelatin-based hemostatic sponge was evaluated by Cenni et al., by assessing its cytotoxicity and genotoxicity (Elisabettacenni et al., 2000). The analysis was carried out by using an extract of the sponge and a continuous cell line, and it was observed that the diluted extract was well tolerated by cells with no toxicity. Genotoxicity tests were carried out to evaluate the effect of foreign gelatin sponge on gene mutation and long-term neoplastic transformation inside body. The sample did not induce any damage to the DNA, indicating that gelatin was nongenotoxic.

6.2.2.3 SILK FIBROIN

Silks are natural fibrous proteins usually produced by insects and spiders and consists of a core structural fibroin protein surrounded by sericin, a type of glue-like proteins which hold fibroin fibers together (Yang et al., 2007). Silk Fibroin contains two polypeptide chains linked by a disulfide bridge. In vivo studies on silk biocompatibility showed adverse immune response only when virgin silk was used. But sutures prepared from only fibroin fibers were found to be inert suggesting that the immune response towards the silk could be attributed to the sericin content. So, to be used as biomaterials, silk fibroin must be separated from sericin which has adverse biological response.

Silk fibroin has been used in various biotechnological and biomedical applications like cell cultures, wound dressing, drug delivery, as sutures, tissue engineering, for their excellent biocompatibility and controllable biodegradability.

Lu et al., carried out a study to investigate the mechanism and control of degradation process of silk fibroin in buffered protease XIV solution (Lu et al., 2011). Previously, it was considered that β-sheet structure was critical to stabilize silk fibroin in aqueous environments but Lu et al., found that silk fibroin films with highest β-sheet content achieved the highest degradation rate in their research. They proposed a new degradation mechanism which revealed that crystal content, hydrophilic interaction, and crystal-non-crystal alternate nanostructures of silk fibroin control its degradation behavior. The hydrophilic blocks of silk fibroin were first degraded and then, hydrophobic crystal blocks became free particles and moved into solution (Figure 6.5).

Yang et al., analyze the biocompatibility of silk fibroin with peripheral nerve tissues and cells (Yang et al., 2007). They cultured rat dorsal root ganglia (DRG) on silk fibroin fibers and Schwann cells from rat sciatic nerves in the silk fibroin extract fluid. After 21 d of culture, they found that axons in bundles were paralleled to the silk fibroin fibers and were wrapped with layers of Schwann cell processes which indicated a good biocompatibility existing between silk fibroin and neuritis or Schwann cells of peripheral nerve tissues. They concluded that silk fibroin supported the cell growth of DRG and facilitated the survival of Schwann cells without exerting any significant cytotoxic effects on cells.

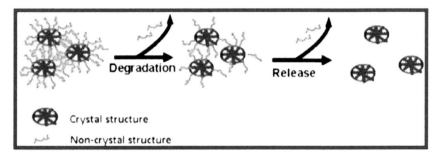

FIGURE 6.5 **(See color insert.)** Degradation mechanism of silk fibroin. Degradation of non-crystal or unstable crystal structures in enzyme solutions which results in the formation of free crystal structure. Then, the crystal structure was dissolved in enzyme solutions. Reprinted with permission from ACS publication (Lu et al., 2011)

6.2.2.4 FIBRIN

Fibrin is a biopolymer of the monomer fibrinogen which is composed of two sets of three polypeptide chains (Aα, Bβ, and γ) bound by six disulfide bridges (De la Puente and Ludeña, 2014). Fibrin is naturally formed after thrombin-mediated cleavage of fibrinopeptide with subsequent conformational changes and exposure of polymerization sites. To form fibrin network, fibrinogen is converted into fibrin through the action of thrombin and activated blood coagulation factor XIIIa which cross-links γ chains in the fibrin polymer and makes the network resistant to protease degradation (Ahmed et al., 2008).

Fibrin is biocompatible with minimal inflammation and foreign body reaction and can be easily degraded with plasmin-mediated fibrinolysis. Because of its biocompatibility and biodegradability, fibrin has been used clinically as a hemostatic agent in cardiac, liver, and spleen surgery; as a sealant in several clinical applications; to reduce suture vascular and intestinal anastomosis, etc. (Ahmed et al., 2008).

Plasminogen and MMPs secreted from the cells are usually responsible for rapid degradation of the fibrin-based biomaterials inside body (Ahmed et al., 2007). However, biodegradability of fibrin can be controlled by a number of techniques like optimizing pH, crosslinking, modification with a molecule such as polyethylene glycol, addition of protease inhibitors, etc.

6.2.2.5 WHEAT GLUTEN

Wheat gluten is a protein that is derived from the second largest cereal crop wheat and has been widely used for various applications because of its abundance, biodegradability, biocompatibility, etc.

Wheat gluten consists of a mixture of proteins that can be classified into two types according to their polymerization properties: monomeric gliadins, and polymeric glutenins (Balaguer et al., 2011). Glutenin consists of polypeptide sequences linked by inter-molecular disulfide bonds and gliadins are monomeric proteins that form only intramolecular disulfide bonds, and gliadin consists of simple polypeptide chains linked by intramolecular disulfide bonds and able to interact with side chains of the glutenin macropolymer through secondary bonds (Domenek et al., 2004). Glutamine and proline are the dominating amino acids in gliadins and glutenins fractions with small amount of charged amino acids, and cysteine is extremely important to the protein structure and functionality as it is responsible for the inter- and intra-chain disulfide bonds (Wieser, 2007). Besides disulfide bonds, there also exist hydrogen bonds, ionic bonds, and hydrophobic interactions in gluten network structure, which significantly determine the degradation behavior of wheat gluten.

6.2.2.6 SOY PROTEIN

Soy protein is extracted from soybean seeds which is primarily an industrial crop cultivated for oil and protein. Because of some outstanding features like sustainability, abundance, low cost, and functionality, film-forming ability, biodegradability, biocompatibility, etc., soy protein has attracted great interest in the field of research for the development of environment-friendly protein-based biomaterials. Soy protein is basically composed of two globular protein fractions β-conglycinin and glycinin with each fraction consists of 20 different amino acids, including lysine, tyrosine, leucine, glutamic acid, phenylalanine, and aspartic acid, etc. (Kumar et al., 2002). The subunits β-conglycinin and glycinin are associated via hydrophobic, hydrogen bonding, and disulfide bonds which decide the degradation behavior of soy protein. The functional properties of soy protein-based products largely depends on the composition and the structure of protein fractions. Biodegradation of soy protein follows the pathway of enzyme-induced hydrolysis. Soy protein undergoes proteolysis by proteases, such as trypsin, alcalase, papain, etc. to produce peptides of smaller molecular size and less quaternary structure than the original proteins. However proteases preferentially hydrolyze β–conglycinin over glycinin because of their difference in structures; the compact structure of glycinin makes it difficult for protease to act (Kim et al., 1990; Ortiz and An, 2000).

Biodegradation of soy protein can be controlled by using cross-linking agent. González et al., soy protein-based biodegradable films and modified it with different amounts of a naturally occurring cross-linking agent, genipin (González et al., 2011).

The biodegradation of the film was observed by burying the films into the soil under indoor conditions for 33 days. Results revealed that degradation of the different films in soil largely depends on the degree of cross-linking. The non-cross-linked film was almost completely degraded in 14 days, whereas the film with 10% crosslinking with genipin remained unaltered after 33 days (Figure 6.6). This can be explained as in most cross-linked films, microbial attack, proteolytic enzyme action, and hydrolysis occur to a lesser extent.

6.3 BIOCOMPATIBILITY OF NATURAL POLYMERS

Biocompatibility is probably the most important criteria for any polymer to be used in medical devices. No polymeric medical device is used without prior biocompatibility testing. Synthetic polymers with their tailor-made properties are more likely to be biocompatible. The abandoned natural polymers have very good biodegradability and clinical usage; however, they can only be used in medical purposes if they pass the biocompatibility test satisfactorily.

Biocompatibility testing of chitosan by cytotoxicity test did not detect any level of formation of endotoxin and neither produce inflammatory response. The test provided information that neutrophils migrate early in the chitosan embedded implant site, normal granulation of tissue happened along with typical healing progress (VandeVord et al., 2002). It is found in another study that the biocompatibility of chitosan increases with the increase of deacetylation of it (Tomihata and Ikada, 1997). Sandra et al., studied the biocompatibility of starch-polyvinyl alcohol based materials and their composites with HA in both in vitro and in vivo models, the materials did not induce adverse reactions (Mendes et al., 2001). Miyamoto et al., attempted two in vivo tests with regenerated cellulose and cellulose derivatives, one for absorbance by living and tissue and one for foreign body reaction. It was observed that in vivo absorbance depends on the degree of crystallinity, and chemical structure of the samples and foreign body reaction was relatively mild. They had concluded that cellulose can be converted to a biocompatible material (Miyamoto et al., 1989). A study was conducted on the hemocompatibility of silk nanoparticles, using whole human blood under quasi-static and flow conditions. The inflammatory response to silk nanoparticles was significantly low underflow versus quasi-static conditions. Silk nanoparticles also had very low coagulant properties. Moreover, the cytotoxicity test did not show any evidence for the formation of endotoxins (Maitz et al., 2017).

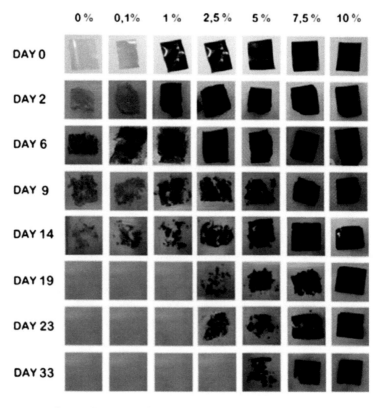

FIGURE 6.6 (See color insert.) Macroscopic appearance of the buried soy protein films with 0%; 0.1%; 1%; 2.5%; 5%; 7.5% and 10% (w/w) of genipin after 0, 2, 6, 9, 14, 19, 23 and 33 days.

Reprinted with permission from Elsevier (González et al., 2011)

Biodegradability and Biocompatibility of Natural Polymers   155

6.3.1   BIOCOMPATIBILITY TESTING METHODS

There are several standards for biocompatibility testing and they are described in ISO 10993, United States Pharmacopeia (USP), European Pharmacopeia (EP), Japanese Ministry of Health, Labor, and Welfare (MHLW) and Japanese Pharmacopeia (JP). Biocompatibility assessment test includes Cytotoxicity, Sensitization, Irritation Reactivity, Systemic Toxicity, Pyrogenicity, Genetic Toxicology, Implantation, Hemocompatibility, however, chemical characterization of the test polymers and their extractable and leachable also help in identifying and selecting suitable biocompatible polymers. These tests are carried on in three groups as termed Level I or primary, level II or secondary and Level III or preclinical. Level I conduct both in vitro and in vivo tests, Level II carries in vivo tests and if any polymer passes both level I and level II tests then the polymer is tested in human body to evaluate the final desirable and adverse reactions of it under normal clinical conditions. In vitro test assesses the effect of direct cell culture activity in the polymers in an experiment. In *in-vivo* tests, the sample material is implanted in subcutaneous or intramuscular (IM) place in rabbit or mouse to observe the tissue reactions.

6.3.1.1   CYTOTOXICITY TESTING

It is a rapid, less expensive and very effective in vitro test which determine the harmful leachable and their effect on cellular components. The test is mainly of three types; e.g., Agar overlay, mammalian cell culture media (MEM) elution and direct contact method. Agar overlay technique consists of culturing a toxic compound onto a solidified soft agar overlay where a bacterial test strain is seeded earlier. In elution method, the test specimen is kept in a growth media (MEM with 5–10% serum) at 37°C for 24 to 72 hours along with a vehicle that will allow extraction of polar and nonpolar components from the sample. For some materials, a direct contact method can be applied to see the surface topographical effect on cytotoxicity.

6.3.1.2   SENSITIZATION

This in vivo test is actually a test for hypersensitivity due to adverse reactions (e.g., symptoms can be redness or swelling) in animals by exposing the animal to the polymer material. The test method can of three types, and they are Kligman Guinea Pig Maximization Test (GPMT), Buehler Sensitization, and Murine Local Lymph Node Assay.

6.3.1.3 IRRITATION/INTRACUTANEOUS REACTIVITY

Irritation tests are in vivo test to assess that whether the test sample causes any irritation on the part of the body after direct exposure. Standard studies are single exposure evaluations and based on the part of the body to be tested, and they are designated as intracutaneous, dermal, oral/ buccal mucosal, ocular, vaginal, rectal, bladder, and penile.

6.3.1.4 SYSTEMIC TOXICITY TESTING

This in vivo test evaluates the potential systemic toxicity of any material to the impairment or activation cells or organs which are remote from the site of contact after different categorical exposures and dose route. Acute systemic toxicity includes a single direct exposure or to an extract with a 72-hour observation period. Repeat or continuous exposure periods are classified as Sub-acute (14 to 28 days), Sub-chronic(90–100 days), and Chronic(longer than 100days), and they comprises total clinical pathology like; clinical chemistry, hematology, and coagulation, necropsy, and organ weights, as well as histopathology. The dose routes include implantation, intraperitoneal, intravenous (IV), oral, and subcutaneous.

6.3.1.5 PYROGEN TESTING

The purpose of the study is to determine the potential presence of pyrogens in extracts of test materials. The in vivo test involves measuring the rise in temperature of rabbits or mouse following the IV injection of a test article extract. There are two sources of pyrogens that should be considered when addressing pyrogenicity. One is material-mediated pyrogens that are chemicals, and they can leach from the material, and another type of pyrogens may be from bacterial endotoxins. It is designed for products that can be tolerated by the test rabbit in a dose not to exceed 10 mL per kg, administered within a period of not more than 10 minutes.

6.3.1.6 GENETIC TOXICOLOGY TESTING

Genotoxicity tests can be conducted either in vivo or in vitro. This test determine the potential of the test article or test article extract to induce gene mutations or chromosome in bacterial, mammalian cells. Physical or chemical agents that induce such effects by interacting with genetic material and alter their structure are called genotoxic. This information is very important as preclinical studies because genetic damage can cause an increase in the incidence of heritable diseases and cancer in human populations (Ren et al., 2017). The test can be done either using prokaryotics (Ames test with *Salmonella typhimurium, Escherichia coli)* *or* using eukaryotics such as chromosomal aberrations with V79 cells or human lymphocytes, Mouse lymphoma assay, Hypoxanthine-guanine Phosphoribosyl Transferase assay, Micronucleus assay in vitro, Micronucleus assay in vivo, DNA-repair: UDS assay in vitro/ex vivo with rat primary hepatocytes.

6.3.1.7 IMPLANTATION TESTING

This in-vivo test assess the local pathological effects at both the gross and microscopic level on living tissues such as muscle, bone, subcutaneous, intraperitoneal, and brain where the sample is implanted. The assay is typically performed in albino rabbits and takes about 4–8 weeks to complete.

6.3.1.8 HEMOCOMPATIBILITY TESTING

Hemocompatibility tests are in vitro assays (Static systems and Dynamic systems) used to assess the possibility of a test article to cause adverse effects on red blood cells (hemolysis), thrombosis, coagulation, platelets, and complement system. The purpose of test is to determine the potential hemolytic effect (destruction of red blood cells with subsequent release of hemoglobin) of the test article extract on blood.

Along with the biocompatibility tests, polymers need some physiochemical test for identification and analysis of their extracts. Many commercial polymers are with additives such as antioxidants, stabilizers, lubricants, plasticizers, colorants, and others. The extracted solution should undergo for some tests for identification of the material, physiochemical properties, and extractable metals. Identification (FTIR, DSC), physicochemical tests (absorbance, acidity/alkalinity, TOC), extractable metals (ICP/MS, ICP/OES), plastic additives, including phenolic and nonphenolic antioxidants (HPLC).

6.4 BIOMEDICAL AND PHARMACEUTICAL APPLICATIONS OF NATURAL POLYMERS

Due to the reasons of biodegradability, nontoxicity, biocompatibility, economic, and easy availability of natural polymers, their uses in biomedical and pharmaceutical industries are more attractive than synthetic polymers. There are many biomedical applications of natural polymers, including bone repair or replacement, dental repair or replacement, tissue engineering, drug delivery with controlled release, skin treatment, etc. For example, collagen has been broadly investigated for numerous medical uses because of its important properties of degradability and biocompatibility as well as mechanical strength. It is used as scaffold or matrix for tissue engineering because it is the key protein component to provide support to various connective tissues of our body (Ramshaw, 2016; Chevallay and Herbage, 2000; Kemp, 2000; Lee et al., 2001). Sponges of collagen have been applied as scaffolds and tissue supports for almost 50 years (Bellucci and Wolff, 1964). Moreover, collagen-based composites have been explored as these materials could mimic the composition of natural bone (Venugopal et al., 2008).

Gelatin is another important protein-based natural polymer. It is safe and it can form polyion complex and these properties are favorable to be used as targeted drug delivery system (tissues of bone, cartilage, skin etc.) (Su and Wang, 2015; Holland et al., 2005; Ito et al., 2003). Commercially available gelatin-based carriers are also used in drug delivery and tissue engineering (Malda et al., 2003). Silks are other types of protein-based natural polymer which have slow degradability (Horan et al., 2005), biocompatibility (Altman et al., 2003). Therefore, silk-based scaffolds or matrices for drug and cell delivery applied in various tissue engineering uses (Li et al., 2006; Wang et al., 2005)

The application of fibrin as a biomaterial has long history which has been revealed to be biodegradable, biocompatible, injectable, and capable to increase cell proliferation (Shaikh et al., 2008). Fibrin glues can be made as solutions having thrombin and fibronectin distinctly, which are mixed just before use. Thrombin quickly crosslinks the fibronectin into a fibrin clot and close the wound. Fibrin also applied as a drug delivery device (Yang et al., 2010) and cell carrier (Dickhut et al., 2009).

On the other hand, polysaccharide-origin polymers also have many biomedical applications as scaffolds or matrices for drug, cell, and gene delivery designated for various tissue engineering uses (Zhang et al., 2006; Torres et al., 2007). Naturally occurring plant-derived polymers are appropriate for the pharmaceutical formulation production in different forms like nanoparticles, implants, films, microparticles, microsphere, blends, matrix, and viscous liquid. Various types of natural polysaccharides are used in the pharmaceutical applications, for example, cellulose, dextrin, alginate, chitosan, Arabic gum, pectin, guar gum, starch, insulin, xanthan gum, carrageenans, etc. (Beneke et al., 2009). In the design of controlled release system, usually biodegradable polymers are more pronounced.

As a polysaccharide, powdered cellulose obtained from wood and cotton has been used in tablets as filler in the pharmaceutical industry. Moreover, microcrystalline cellulose produced from good quality powdered cellulose can be used as diluents or filler/binder in tablets in granulation and or direct compression procedures (Kibbe, 2000). Furthermore, cellulose derivatives can be used in membrane controlled discharge systems like enteric coating and the application of semi-permeable membranes in osmotic pump delivery processes. These derivatives also have capability to sustain the release of drugs and they are using for controlled drug release system (Andreopoulos and Tarantili, 2002; Conti et al., 2007).

As a natural polysaccharide, chitosan is one of the most important polymer which has been demonstrated for the delivery of many biopharmaceutical like human growth hormone (Cheng et al., 2005), insulin (Chaudhury and Das, 2011), genetic material, DNA, vaccines (Beneke et al., 2009), and condensed plasmid DNA (Bowman and Leong, 2006), etc. In other words, chitosan, with its exciting properties, is one of the most encouraging natural polymers for tissue engineering, drug delivery, theranostics, and gene therapy (Dash et al., 2011). Chitosan is also used as a coating material over the core tablet which can protect the drug from being released in the physiological atmosphere of stomach and small intestine and is vulnerable to colonic bacterial enzymatic activities with subsequent drug release in the colon (Ogaji et al., 2012).

Starch, as a native or modified form has been using as one of the main excipients in the formulation of pharmaceutical tablet and capsule. It is also used as binder, bulking, disintegrant or aiding drug delivery. Starch derivative (e.g., starch-acetate) systems have been reported for controlled drug delivery (Tuovinen et al., 2003). Moreover, native starches and amylose have been used as film-forming material in pharmaceutical film coatings and a mixture of amylose and ethyl-cellulose (aqueous and non-aqueous) based coatings for drug delivery in colon (Milojevic et al., 1996; Palviainen et al., 2001)

Alginates (kind of natural biodegradable polysaccharide) have numerous uses in drug delivery (e.g., matrix type alginate gel beads) (Tønnesen and Karlsen, 2002). Alginates have extensive applications in the biomedical and biopharmaceutical fields, where there are controlled delivery systems like vaccines (Beneke et al., 2009), insulin, and growth hormones (Beneke et al., 2009). It is also applied for the delivery of somatic gene therapy by producing the formulation of microcapsule.

Gums have various applications as emulsifiers, sweeteners, viscosi-fiers, thickeners, etc. in confectionary. In pharmaceutical industry, they have use as binders and modifiers for drug release. Their applications as drug delivery is lower because of the lack of functional groups on the backbone of natural gums. Therefore, grafting of copolymers onto gums (gum-g-copolymers) show more advantages in this area due to the incorporation of more functional groups which have the capability to load drugs. For example, when AG-g-PAM copolymer was prepared by grafting of acrylamide onto guar gum, controlled release of diltiazem hydrochloride and nifedipine was happened (Toti and Aminabhavi, 2004). Again, *N*-isopropylacrylamide-grafted-acacia gum polymer produced temperature-sensitive drug delivery (Kalia et al., 2013).

6.5 CONCLUDING REMARKS

This chapter presented the types and properties of natural polymers based on biodegradability and biocompatibility and finally their biomedical and pharmaceutical applications. The natural-based polymer properties for the aimed uses were explained with a distinct emphasis in the clari-fication of protein-origin polymers and polysaccharides because of their resemblance with the ECM. Polymers from natural sources have some disadvantages like the problems in regulating the variability from batch to batch and mechanical properties or restricted processability however, their benefits obviously surplus the limitations. Their biocompatibility, biodegradability, low price and obtainability, resemblance with the ECM creates them very potential candidates in the biomedical field of applications. However, for biodegradable materials, long-term tests are necessary to allow the assessment of the effects of constant release of metabolites causing from their degradation and to detect the kind and degree of local and systemic variations. Therefore, the combination of numerous *in vitro* and *in vivo* tests can offer a summary of the interaction of biomaterials with the host. Again, for a material to be biocompatible, it must be subjected to many studies extending from *in vitro* assays to clinical trials and including different areas like chemistry, pharmaceutics, toxicology, and biology. From the analysis of different standard tests result, one can understand the properties of the materials including their safety in relation to cells and tissues. Therefore, the interaction of biomaterials with the host can only be obtained by a combined analysis of different *in vitro* and *in vivo* tests. Another important field of applications of natural-based polymers are as pharmaceutical excipients which have been continuing to lead the research efforts of scientists to find cheaper, biodegradable, and sustainable excipients. Many of these excipients have clear benefits over the synthetic one in some specific delivery systems because of their intrinsic properties, and their applications will be increased in the future.

KEYWORDS

- **biocompatibility**
- **biodegradability**
- **biomaterial**
- **controlled release**
- **drug delivery**
- **excipient**
- **natural polymer**

REFERENCES

Ahmed, T. A., Dare, E. V., & Hincke, M., (2008). Fibrin: A versatile scaffold for tissue engineering applications. *Tissue Eng. Part. B Rev.*, *14*(2), 199–215.

Ahmed, T. A., Griffith, M., & Hincke, M., (2007). Characterization and inhibition of fibrin hydrogel–degrading enzymes during development of tissue engineering scaffolds. *Tissue Eng.*, *13*(7), 1469–1477.

Aimes, R. T., & Quigley, J. P., (1995). Matrix metalloproteinase-2 is an interstitial collagenase inhibitor-free enzyme catalyzes the cleavage of collagen fibrils and soluble native type I collagen generating the specific ¾-and ¼-length fragments. *J. Biol. Chem.*, *270*(11), 5872–5876.

Altman, G. H., Diaz, F., Jakuba, C., Calabro, T., Horan, R. L., Chen, J., Lu, H., Richmond, J., & Kaplan, D. L., (2003). Silk-based biomaterials. *Biomaterials*, *24*(3), 401–416.

Amass, W., Amass, A., & Tighe, B., (1998). A review of biodegradable polymers: Uses, current developments in the synthesis and characterization of biodegradable polyesters, blends of biodegradable polymers and recent advances in biodegradation studies. *Polym. Int.*, *47*(2), 89–144.

Andreopoulos, A., & Tarantili, P., (2002). Study of biopolymers as carriers for controlled release. *J. Macromol. Sci. Part B Phys.*, *41*(3), 559–578.

Balaguer, M. P., Gómez-Estaca, J., Gavara, R., & Hernandez-Munoz, P., (2011). Biochemical properties of bioplastics made from wheat gliadins cross-linked with cinnamaldehyde. *J. Agric. Food Chem.*, *59*(24), 13212–13220.

Baran, E., Mano, J., & Reis, R., (2004). Starch–chitosan hydrogels prepared by reductive alkylation cross-linking. *J. Mater. Sci. Mater. Med.*, *15*(7), 759–765.

Bassleer, C., Combal, J. P. A., Bougaret, S., & Malaise, M., (1998). Effects of chondroitin sulfate and interleukin-1β on human articular chondrocytes cultivated in clusters. *Osteoarthr. Cartil.*, *6*(3), 196–204.

Bellucci, R., & Wolff, D., (1964). Experimental stapedectomy with collagen sponge implant. *Laryngoscope*, *74*(5), 668–688.

Beneke, C. E., Viljoen, A. M., & Hamman, J. H., (2009). Polymeric plant-derived excipients in drug delivery. *Molecules*, *14*(7), 2602–2620.

Bowman, K., & Leong, K. W., (2006). Chitosan nanoparticles for oral drug and gene delivery. *Int. J. Nanomed.*, *1*(2), 117.

Campo, V. L., Kawano, D. F., Da Silva, D. B., & Carvalho, I., (2009). Carrageenans: Biological properties, chemical modifications and structural analysis: A review. *Carbohydrate Polymers*, *77*(2), 167–180.

Cascone, M., Barbani, N., P. Giusti, C. C., Ciardelli, G., & Lazzeri, L., (2001). Bioartificial polymeric materials based on polysaccharides. *J. Biomater. Sci., Polym. Ed.*, *12*(3), 267–281.

Chandra, R., & Rustgi, R., (1998). Biodegradable polymers. *Prog. Polym. Sci.*, *23*(7), 1273–1335.

Chasin, M., Lewis, D., & Langer, R., (1988). Polyanhydrides for controlled drug delivery. *Biopharm. Manufact.*, *1*, 33–46.

Chatakanonda, P., Varavinit, S., & Chinachoti, P., (2000). Effect of crosslinking on thermal and microscopic transitions of rice starch. *LWT – Food Sci. Technol.*, *33*(4), 276–284.

Chaudhury, A., & Das, S., (2011). Recent advancement of chitosan-based nanoparticles for oral controlled delivery of insulin and other therapeutic agents. *Aaps Pharmscitech*, *12*(1), 10–20.

Cheng, Y. H., Dyer, A. M., Jabbal-Gill, I., Hinchcliffe, M., Nankervis, R., Smith, A., & Watts, P., (2005). Intranasal delivery of recombinant human growth hormone (somatropin) in sheep using chitosan-based powder formulations. *Eur. J. Pharm. Sci.*, *26*(1), 9–15.

Chevallay, B., & Herbage, D., (2000). Collagen-based biomaterials as 3D scaffold for cell cultures: Applications for tissue engineering and gene therapy. *Med. Biol. Eng. Comput.*, *38*(2), 211–218.

Choi, S., Choi, W., Kim, S., Lee, S. Y., Noh, I., & Kim, C. W., (2014). Purification and biocompatibility of fermented hyaluronic acid for its applications to biomaterials. *Biomater. Res.*, *18*(1), 6.

Cohen, S., Yoshioka, T., Lucarelli, M., Hwang, L. H., & Langer, R., (1991). Controlled delivery systems for proteins based on poly (lactic/glycolic acid) microspheres. *Pharm. Res.*, *8*(6), 713–720.

Colombo, P., Bettini, R., & Peppas, N. A., (1999). Observation of swelling process and diffusion front position during swelling in hydroxypropyl methylcellulose (HPMC) matrices containing a soluble drug. *J. Control. Release*, *61*(1/2), 83–91.

Conti, S., Maggi, L., Segale, L., Machiste, E. O., Conte, U., Grenier, P., & Vergnault, G., (2007). Matrices containing NaCMC and HPMC: 1. Dissolution performance characterization. *Int. J. Pharm. Pharm. Sci.*, *333*(1/2), 136–142.

Danisovic, L., Bohac, M., Kuniakova, M., Csobonyeiova, M., Oravcova, L., Novakova, V. Z., & Bohmer, D., (2013). In vitro testing of modified collagen/hyaluronan/β-glucan scaffold for tissue engineering application. *Regen. Med.*, *8*(6), 179.

Dash, M., Chiellini, F., Ottenbrite, R., & Chiellini, E., (2011). Chitosan—A versatile semi-synthetic polymer in biomedical applications. *Prog. Polym. Sci.*, *36*(8), 981–1014.

De la Puente, P., & Ludeña, D., (2014). Cell culture in autologous fibrin scaffolds for applications in tissue engineering. *Exp. Cell Res.*, *322*(1), 1–11.

Dickhut, A., Dexheimer, V., Martin, K., Lauinger, R., Heisel, C., & Richter, W., (2009). Chondrogenesis of human mesenchymal stem cells by local transforming growth factor-beta delivery in a biphasic resorbable carrier. *Tissue Eng. Part A*, *16*(2), 453–464.

Domenek, S., Brendel, L., Morel, M. H., & Guilbert, S., (2004). Swelling behavior and structural characteristics of wheat gluten polypeptide films. *Biomacromolecules*, *5*(3), 1002–1008.

Draget, K. I., & Taylor, C., (2011). Chemical, physical and biological properties of alginates and their biomedical implications. *Food Hydrocoll.*, *25*(2), 251–256.

Duceppe, N., & Tabrizian, M., (2010). Advances in using chitosan-based nanoparticles for in vitro and in vivo drug and gene delivery. *Expert Opin. Drug. Deliv.*, *7*(10), 1191–1207.

Elisabettacenni, C. G., Stea, S., Corradini, A., & Carozzi, F., (2000). Biocompatibility and performance *in vitro* of a hemostatic gelatin sponge. *J. Biomater. Sci. Polym. Ed.*, *11*(7), 685–699.

González, A., Strumia, M. C., & Igarzabal, C. I. A., (2011). Cross-linked soy protein as material for biodegradable films: Synthesis, characterization and biodegradation. *J. Food Eng.*, *106*(4), 331–338.

Gorgieva, S., & Kokol, V., (2011). Collagen vs. gelatine-based biomaterials and their biocompatibility: Review and perspectives. In: *Biomaterials Applications for Nanomedicine*, InTech., vol. 2, pp. 17–52.

Harkness, R., (1966). Collagen. *Sci. Prog.*, *54*(214), 257–274.

Holland, T. A., Tabata, Y., & Mikos, A. G., (2005). Dual growth factor delivery from degradable oligo (poly (ethylene glycol) fumarate) hydrogel scaffolds for cartilage tissue engineering. *J. Control. Release*, *101*(1–3), 111–125.

Horan, R. L., Antle, K., Collette, A. L., Wang, Y., Huang, J., Moreau, J. E., Volloch, V., Kaplan, D. L., & Altman, G. H., (2005). *In vitro* degradation of silk fibroin. *Biomaterials*, *26*(17), 3385–3393.

Ito, A., Mase, A., Takizawa, Y., Shinkai, M., Honda, H., Hata, K. I., Ueda, M., & Kobayashi, T., (2003). Transglutaminase-mediated gelatin matrices incorporating cell adhesion factors as a biomaterial for tissue engineering. *J. Biosci. Bioeng.*, *95*(2), 196–199.

Izydorczyk, M., Cui, S. W., & Wang, Q., (2005). *Polysaccharide Gums: Structures, Functional Properties, and Applications* (pp. 263–307). Taylor & Francis.

Jayakumar, R., Menon, D., Manzoor, K., Nair, S., & Tamura, H., (2010). Biomedical applications of chitin and chitosan-based nanomaterials: A short review. *Carbohydr. Polym.*, *82*(2), 227–232.

Kalia, S., Sabaa, M. W., & Kango, S., (2013). Polymer grafting: A versatile means to modify the polysaccharides. In: *Polysaccharide-Based Graft Copolymers* (pp. 1–14). Springer.

Kemp, P. D., (2000). *Tissue Engineering and Cell-Populated Collagen Matrices* (Vol. 139, pp. 287–293).

Kibbe, A., (2000). *Handbook of Pharmaceutical Excipients* (Vol. 160, pp. 276–278, 324). Pharmaceutical Press London, United Kingdom and American Pharmaceutical Association, Washington, DC.

Kim, S. K., Byun, H. G., Park, P. J., & Shahidi, F., (2001). Angiotensin I converting enzyme inhibitory peptides purified from bovine skin gelatin hydrolysate. *J. Agric. Food Chem.*, *49*(6), 2992–2997.

Kim, S. Y., Park, P. S., & Rhee, K. C., (1990). Functional properties of proteolytic enzyme modified soy protein isolate. *J. Agric. Food Chem.*, *38*(3), 651–656.

Klemm, D., Heublein, B., Fink, H. P., & Bohn, A., (2005). Cellulose: Fascinating biopolymer and sustainable raw material. *Angew. Chem. Int. Ed.*, *44*(22), 3358–3393.

Kumar, A., Srivastava, A., Galaev, I. Y., & Mattiasson, B., (2007). Smart polymers: Physical forms and bioengineering applications. *Prog. Polym. Sci.*, *32*(10), 1205–1237.

Kumar, R., Choudhary, V., Mishra, S., Varma, I., & Mattiason, B., (2002). Adhesives and plastics based on soy protein products. *Ind. Crops Prod.*, *16*(3), 155–172.

Kusumocahyo, S. P., Ichikawa, T., Shinbo, T., Iwatsubo, T., Kameda, M., Ohi, K., Yoshimi, Y., & Kanamori, T., (2005). Pervaporative separation of organic mixtures using dinitrophenyl group-containing cellulose acetate membrane. *J. Membr. Sci.*, *253*(1/2), 43–48.

Laurent, T. C., Laurent, U., & Fraser, J., (1995). Functions of hyaluronan. *Ann. Rheum. Dis.*, *54*(5), 429.

Lee, C. H., Singla, A., & Lee, Y., (2001). Biomedical applications of collagen. *Int. J. Pharm. Pharm. Sci.*, *221*(1/2), 1–22.

Lee, C. T., Kung, P. H., & Lee, Y. D., (2005). Preparation of poly (vinyl alcohol)-chondroitin sulfate hydrogel as matrices in tissue engineering. *Carbohydr. Polym.*, *61*(3), 348–354.

Lee, K. Y., & Mooney, D. J., (2012). Alginate: Properties and biomedical applications. *Prog. Polym. Sci., 37*(1), 106–126.

Li, C., Vepari, C., Jin, H. J., Kim, H. J., & Kaplan, D. L., (2006). Electrospun silk-BMP-2 scaffolds for bone tissue engineering. *Biomaterials, 27*(16), 3115–3124.

Liao, Y. H., Jones, S. A., Forbes, B., Martin, G. P., & Brown, M. B., (2005). Hyaluronan: Pharmaceutical characterization and drug delivery. *Drug Deliv., 12*(6), 327–342.

Lu, Q., Zhang, B., Li, M., Zuo, B., Kaplan, D. L., Huang, Y., & Zhu, H., (2011). Degradation mechanism and control of silk fibroin. *Biomacromolecules, 12*(4), 1080–1086.

Maitz, M. F., Sperling, C., Wongpinyochit, T., Herklotz, M., Werner, C., & Seib, F. P., (2017). Biocompatibility assessment of silk nanoparticles: Hemocompatibility and internalization by human blood cells. *Nanomed. Nanotechnol. Biol. Med., 13*(8), 2633–2642.

Malafaya, P. B., Silva, G. A., & Reis, R. L., (2007). Natural–origin polymers as carriers and scaffolds for biomolecules and cell delivery in tissue engineering applications. *Adv. Drug Deliv. Rev., 59*(4/5), 207–233.

Malda, J., Kreijveld, E., Temenoff, J. S., Van Blitterswijk, C. A., & Riesle, J., (2003). Expansion of human nasal chondrocytes on macroporous microcarriers enhances redifferentiation. *Biomaterials, 24*(28), 5153–5161.

Malhotra, B., Keshwani, A., & Kharkwal, H., (2015). Natural polymer based cling films for food packaging. *Int. J. Pharm. Pharm. Sci., 7*, 10–18.

Mallik, A. K., Rahman, M. M., & Ihara, H., (2018). Peptide-based derivative-grafted silica for molecular recognition system: Synthesis and characterization. In: *Biopolymer Grafting* (pp. 235–294). Elsevier.

Malviya, R., Srivastava, P., & Kulkarni, G., (2011). Applications of mucilages in drug delivery-a review. *Adv. Biol. Res., 5*(1), 1–7.

Mangione, M., Giacomazza, D., Bulone, D., Martorana, V., Cavallaro, G., & San Biagio, P., (2005). K+ and Na+ effects on the gelation properties of κ-carrageenan. *Biophys. Chem., 113*(2), 129–135.

Marques, A., Reis, R., & Hunt, J., (2002). The biocompatibility of novel starch-based polymers and composites: *In vitro* studies. *Biomaterials, 23*(6), 1471–1478.

Mendes, S. C., Reis, R., Bovell, Y. P., Cunha, A., Van Blitterswijk, C. A., & De Bruijn, J. D., (2001). Biocompatibility testing of novel starch-based materials with potential application in orthopedic surgery: A preliminary study. *Biomaterials, 22*(14), 2057–2064.

Meyer, D. E., & Chilkoti, A., (2002). Genetically encoded synthesis of protein-based polymers with precisely specified molecular weight and sequence by recursive directional ligation: Examples from the elastin-like polypeptide system. *Biomacromolecules, 3*(2), 357–367.

Meyers, M. A., Chen, P. Y., Lin, A. Y. M., & Seki, Y., (2008). Biological materials: Structure and mechanical properties. *Prog. Mater. Sci., 53*(1), 1–206.

Milojevic, S., Newton, J. M., Cummings, J. H., Gibson, G. R., Botham, R. L., Ring, S. G., Stockham, M., & Allwood, M. C., (1996). Amylose as a coating for drug delivery to the colon: Preparation and *in vitro* evaluation using 5-aminosalicylic acid pellets. *J. Control. Release, 38*(1), 75–84.

Miyamoto, T., Takahashi, S. I., Ito, H., Inagaki, H., & Noishiki, Y., (1989). Tissue biocompatibility of cellulose and its derivatives. *J. Biomed. Mater. Res. Part A, 23*(1), 125–133.

Naessens, M., Cerdobbel, A., Soetaert, W., & Vandamme, E. J., (2005). Leuconostoc dextransucrase and dextran: production, properties and applications. *J. Chem. Technol. Biotechnol.*, *80*(8), 845–860.

Nair, L. S., & Laurencin, C. T., (2007). Biodegradable polymers as biomaterials. *Prog. Polym. Sci.*, *32*(8/9), 762–798.

Nishinari, K., & Takahashi, R., (2003). Interaction in polysaccharide solutions and gels. *Current Opinion in Colloid & Interface Science*, *8*(4/5), 396–400.

Ogaji, I. J., Nep, E. I., & Audu-Peter, J. D., (2012). *Advances in Natural Polymers as Pharmaceutical Excipients*, *3*(1), 1–16.

Orive, G., Carcaboso, A., Hernandez, R., Gascon, A., & Pedraz, J., (2005). Biocompatibility evaluation of different alginates and alginate-based microcapsules. *Biomacromolecules*, *6*(2), 927–931.

Ortiz, S. E. M., & An, M. C., (2000). Analysis of products, mechanisms of reaction, and some functional properties of soy protein hydrolysates. *J. Am. Oil Chem.' Soc.*, *77*(12), 1293–1301.

Palviainen, P., Heinämäki, J., Myllärinen, P., Lahtinen, R., Yliruusi, J., & Forssell, P., (2001). Corn starches as film formers in aqueous-based film coating. *Pharm. Dev. Technol.*, *6*(3), 353–361.

Park, H., & Park, K., (1996). Biocompatibility issues of implantable drug delivery systems. *Pharm. Res.*, *13*(12), 1770–1776.

Park, Y. J., Lee, Y. M., Lee, J. Y., Seol, Y. J., Chung, C. P., & Lee, S. J., (2000). Controlled release of platelet-derived growth factor-BB from chondroitin sulfate–chitosan sponge for guided bone regeneration. *J. Control. Release*, *67*(2/3), 385–394.

Perepelkin, K., (2005). Polymeric materials of the future based on renewable plant resources and biotechnologies: Fibers, films, plastics. *Fiber Chem.*, *37*(6), 417–430.

Piecuch, J. F., & Fedorka, N. J., (1983). Results of soft-tissue surgery over implanted replamine form hydroxyapatite. *J. Oral Maxillofac. Surg.*, *41*(12), 801–806.

Pieper, J., Oosterhof, A., Dijkstra, P. J., Veerkamp, J., & Van Kuppevelt, T., (1999). Preparation and characterization of porous crosslinked collagenous matrices containing bioavailable chondroitin sulfate. *Biomaterials*, *20*(9), 847–858.

Postlethwaite, A. E., Seyer, J. M., & Kang, A. H., (1978). Chemotactic attraction of human fibroblasts to type I, II, and III collagens and collagen-derived peptides. *Proc. Natl. Acad. Sci. U.S.A.*, *75*(2), 871–875.

Prabhanjan, H., Gharia, M., & Srivastava, H., (1989). Guar gum derivatives. Part I: Preparation and properties. *Carbohydr. Polym.*, *11*(4), 279–292.

Rahmanian, S. A., Held, M., Knoeller, T., Stachon, S., Schmidt, T., Schaller, H. E., & Just, L., (2014). *In vivo* biocompatibility and biodegradation of a novel thin and mechanically stable collagen scaffold. *J. Biomed. Mater. Res. Part A*, *102*(4), 1173–1179.

Ramshaw, J. A., (2016). Biomedical applications of collagens. *J. Biomed. Mater. Res. B*, *104*(4), 665–675.

Ratner, B. D., Horbett, T., Hoffman, A. S., & Hauschka, S. D., (1975). Cell adhesion to polymeric materials: Implications with respect to biocompatibility. *J. Biomed. Mater. Res. Part A*, *9*(5), 407–422.

Ren, N., Atyah, M., Chen, W. Y., & Zhou, C. H., (2017). The various aspects of genetic and epigenetic toxicology: Testing methods and clinical applications. *J. Transl. Med.*, *15*(1), 110.

Rottensteiner, U., Sarker, B., Heusinger, D., Dafinova, D., Rath, S. N., Beier, J. P., Kneser, U., Horch, R. E., Detsch, R., & Boccaccini, A. R., (2014). *In vitro* and *in vivo* biocompatibility of alginate dialdehyde/gelatin hydrogels with and without nanoscaled bioactive glass for bone tissue engineering applications. *Materials*, *7*(3), 1957–1974.

Salgado, A., Coutinho, O., Reis, R., & Davies, J., (2007). *In vivo* response to starch-based scaffolds designed for bone tissue engineering applications. *J. Biomed. Mater. Res. Part A*, *80*(4), 983–989.

Satturwar, P. M., Fulzele, S. V., & Dorle, A. K., (2003). Biodegradation and *in vivo* biocompatibility of rosin: A natural film-forming polymer. *AAPS Pharm. Sci. Tech.*, *4*(4), 434–439.

Sezer, A. D., & Cevher, E., (2012). Topical drug delivery using chitosan nano-and microparticles. *Expert Opin. Drug. Deliv.*, *9*(9), 1129–1146.

Shaikh, F. M., Callanan, A., Kavanagh, E. G., Burke, P. E., Grace, P. A., & McGloughlin, T. M., (2008). Fibrin: A natural biodegradable scaffold in vascular tissue engineering. *Cells Tissues Organs*, *188*(4), 333–346.

Shi, C., Zhu, Y., Ran, X., Wang, M., Su, Y., & Cheng, T., (2006). Therapeutic potential of chitosan and its derivatives in regenerative medicine. *J. Surg. Res.*, *133*(2), 185–192.

Silva, G. A., Marques, A., Gomes, M. E., Coutinho, O., & Reis, R. L., (2004). *Cytotoxicity Screening of Biodegradable Polymeric Systems*, *19*, 339–349.

Song, E., Kim, S. Y., Chun, T., Byun, H. J., & Lee, Y. M., (2006). Collagen scaffolds derived from a marine source and their biocompatibility. *Biomaterials*, *27*(15), 2951–2961.

Stephen, A. M., Churms, S., & Vogt, D., (1990). Exudate gums. In: *Methods in Plant Biochemistry* (Vol. 2, pp. 483–522). Elsevier.

Su, K., & Wang, C., (2015). Recent advances in the use of gelatin in biomedical research. *Biotechnol. Lett.*, *37*(11), 2139–2145.

Tomihata, K., & Ikada, Y., (1997). *In vitro* and *in vivo* degradation of films of chitin and its deacetylated derivatives. *Biomaterials*, *18*(7), 567–575.

Tønnesen, H. H., & Karlsen, J., (2002). Alginate in drug delivery systems. *Drug Dev. Ind. Pharm.*, *28*(6), 621–630.

Torres, F. G., Boccaccini, A. R., & Troncoso, O. P., (2007). Microwave processing of starch-based porous structures for tissue engineering scaffolds. *J. Appl. Polym. Sci.*, *103*(2), 1332–1339.

Toti, U. S., & Aminabhavi, T. M., (2004). Modified guar gum matrix tablet for controlled release of diltiazem hydrochloride. *J. Control. Release*, *95*(3), 567–577.

Tuovinen, L., Peltonen, S., & Järvinen, K., (2003). Drug release from starch-acetate films. *J. Control. Release*, *91*(3), 345–354.

VandeVord, P. J., Matthew, H. W., DeSilva, S. P., Mayton, L., Wu, B., & Wooley, P. H., (2002). Evaluation of the biocompatibility of a chitosan scaffold in mice. *J. Biomed. Mater. Res. Part A.*, *59*(3), 585–590.

Venugopal, J., & Ramakrishna, S., (2005). Applications of polymer nanofibers in biomedicine and biotechnology. *Appl. Biochem. Biotechnol.*, *125*(3), 147–157.

Venugopal, J., Low, S., Choon, A. T., Kumar, T. S., & Ramakrishna, S., (2008). Mineralization of osteoblasts with electrospun collagen/hydroxyapatite nanofibers. *J. Mater. Sci. Mater. Med.*, *19*(5), 2039–2046.

Voet, D., Voet, J. G., & Pratt, C. W., (1999). *Fundamentals of Biochemistry* (vol. 1452). Wiley New York.

Wang, Y., Kim, U. J., Blasioli, D. J., Kim, H. J., & Kaplan, D. L., (2005). *In vitro* cartilage tissue engineering with 3D porous aqueous-derived silk scaffolds and mesenchymal stem cells. *Biomaterials*, *26*(34), 7082–7094.

Wieser, H., (2007). Chemistry of gluten proteins. *Food Microbiol.*, *24*(2), 115–119.

Williams, D. F., (2009). On the nature of biomaterials. *Biomaterials*, *30*(30), 5897–5909.

Wilson, Jr. O. C., Blair, E., Kennedy, S., Rivera, G., & Mehl, P., (2008). Surface modification of magnetic nanoparticles with oleylamine and gum Arabic. *Mater. Sci. Eng. C.*, *28*(3), 438–442.

Yang, H. S., Bhang, S. H., Hwang, J. W., Kim, D. I., & Kim, B. S., (2010). Delivery of basic fibroblast growth factor using heparin-conjugated fibrin for therapeutic angiogenesis. *Tissue Eng. Part A*, *16*(6), 2113–2119.

Yang, Y., Chen, X., Ding, F., Zhang, P., Liu, J., & Gu, X., (2007). Biocompatibility evaluation of silk fibroin with peripheral nerve tissues and cells *in vitro*. *Biomaterials*, *28*(9), 1643–1652.

Yannas, I., Burke, J., Orgill, D., & Skrabut, E., (1982). Wound tissue can utilize a polymeric template to synthesize a functional extension of skin. *Science*, *215*(4529), 174–176.

Zhang, Y., Cheng, X., Wang, J., Wang, Y., Shi, B., Huang, C., Yang, X., & Liu, T., (2006). Novel chitosan/collagen scaffold containing transforming growth factor-β1 DNA for periodontal tissue engineering. *Biochem. Biophys. Res. Commun.*, *344*(1), 362–369.

Index

1

1-deamino-8-D-arginine vasopressin, 12
1-ethyl-3-(3-dimethylamino propyl) carbodiimide hydrochloride (EDC), 96, 148
1H nuclear magnetic resonance (1H NMR), 122

3

3-α polypeptide chain, 62

4

4-(p-iodophenyl) butanoic acid derivatives, 41

5

5-fluorouracil, 7, 8

α

α-helical reactive loop, 37
α-helices, 35
α(1→4)
 bonds, 140
 linked glucose residues, 140
α(1→6)
 branch points, 140
 linked D-glucose residues, 146
α-1,4-D-glucuronic acid, 144
α-fetoprotein, 48
α-helix, 39
α-l-guluronic acid, 142

β

β-(1–4) linked D-glucosamine, 138
β-(1→3) and β(1→6)-linked D-galactose units, 145
β-(1→4) and β-(1→3) glycosidic bonds, 1, 5
β-(1→6)-linked D-glucopyranosyl uronic acid units, 145
β-1,3-*N*-acetyl-D-glucosamine, 144
β-conglycinin, 152
β-cyclodextrins films, 11
β-d-glucuronidase, 1, 6
β-D-mannuronic acid, 142
β-N-acetylhexosaminidase, 1, 6
β-sheet, 37, 150
β-tricalcium phosphate, 77

A

Acacia gum, 145, 160
Acetylation, 139
Acid
 base balance, 44
 glucuronide metabolite, 36
 soluble collagen, 67
Acyclovir-loaded albumin nanoparticles, 47
Adipose-derived stem cells, 70
Albendazole, 48, 50
Albondin, 38
Albumin, 3, 18, 33–36, 38–45, 47, 48, 50, 51, 125
 administration, 50
 drug transport, 34, 44
 encapsulation, 33
 nanoparticles, 44, 45, 47, 48, 50
 receptors, 38, 43
 stabilized nanoparticle formulation, 45
 therapy, 51
 types/structure/binding sites, 35
Alcalase, 149, 153
Algae species, 142
Alginate, 103, 142, 143
 dialdehydegelatin (ADA-GEL), 143
Alpha chain, 64
Alport syndrome, 62

Amino
acid, 15, 34, 35, 63, 118, 134, 147, 149, 152
residues, 34–36, 149
sequence, 147
hydroxyl groups, 140
Amylopectin, 140, 141
Angiogenesis, 71, 84, 85, 139
Anti-angiogenic drug, 50
Anti-apoptotic effects, 44
Antibacterial films, 142
Antibody
fragments, 43, 44
secretions, 148
Anticancer
drugs, 14, 48, 117, 123, 128
property, 69
Anticoagulant effect, 44
Anti-epidermal growth factor receptor (EGFR), 42
Anti-inflammatory, 15, 44, 51, 119
Antioxidant, 6, 35, 44, 51, 69, 157, 158
Antitumor drugs, 15
Antiviral drug, 50
Arabinogalactan, 145
Aromatic compound, 35
Arthropods, 138
Ascophyllum nodosum, 142
Ascorbic acid, 64
Aspergillus niger, 98, 103
Autoimmune disorders, 62
Axial hydrogen atoms, 5
Azidothymidine, 50

B

Bacillus subtilis, 4
Bacterial
infection, 107
rhinosinusitis, 13
test strain, 155
Bacteriostatic property, 6
Basement membrane (BM), 62, 63, 84
Basolateral membrane, 38
Benzodiazepine, 35, 51
Bioactive compounds, 33–35, 47
Bioadhesive transdermal device, 9

Bioavailability, 10, 12, 13, 41, 47, 96
Biocompatibility, 2, 3, 7, 48, 51, 61, 77, 79, 80, 93, 104, 118, 133–137, 139, 141–153, 155, 157, 158, 160, 161
assessment test, 155
natural polymers, 136, 153
testing methods, 155
cytotoxicity testing, 155
genetic toxicology testing, 157
hemocompatibility testing, 157
implantation testing, 157
irritation/intracutaneous reactivity, 156
pyrogen testing, 156
sensitization, 155
systemic toxicity testing, 156
Biocompatible polymers, 155
Biodegradability, 2, 7, 14, 61, 64, 76, 118, 133–136, 138, 139, 141, 145–147, 150–153, 158, 160, 161
Biodegradable, 19, 33, 34, 37, 50, 72, 93, 94, 106, 109, 117, 128, 133, 134, 136, 137, 146, 158–161
collagen matrix, 72
natural polymers, 135, 136
Biodegradation, 118, 138, 147–149, 153
Bioflavonoid, 69
Biological activity, 102, 118, 133, 147
Biomaterial, 14, 61, 63, 69, 71, 85, 77, 133–136, 142, 143, 147, 148, 150–152, 158, 161
Biomedical
applications, 104, 107, 141, 150, 158, 159
bioadhesives, 106
fibers, 107
hydrogels, 105
field, 2, 99, 108, 161
pharmaceutical applications, 68, 158
bone substitutes, 76
cancer therapy, 83
cardiology,81
implantable biomaterials, 71
skin replacement, 74
stents/vascular graft coatings, 79
wound dressing, 68
Biopolymers, 2, 3, 16, 93, 109, 134, 141

Index 171

Bone
  grafting, 62
  mineralization, 107
  substitute, 77, 78, 85, 86
Bovine
  serum albumin (BSA), 33, 34, 37, 38, 47, 48, 50, 125
  spongiform encephalopathy (BSE), 95
Breast
  adenocarcinoma cell lines (MDA-MB-231), 124
  cancer, 16, 45, 48, 99, 102
  biopsy fragments, 83
Bupivacaine, 9, 106

C

Camptothecin, 9, 100
Cancer therapy, 61, 83, 142
Candida albicans, 99
Carbodiimide, 106, 148, 149
Carboplatin, 45
Carboxymethyl cellulose (CMC), 8, 138
Carboxy-terminal telopeptide of type-I collagen (ICTP), 82
Carotid endarterectomy, 80
Cationic lipid complexes, 99
Caveolae vesicle, 38
Caveolin-1, 38
Cell
  differentiation, 77, 81, 84
  membrane, 4
  proliferation, 39, 70, 76, 84, 105, 144, 158
Cellulose acetate, 137, 138, 142
  butyrate, 138
  derivatives, 136, 137
  ethers, 137, 138
  trimelitate, 138
Cetuximab, 42
Chitooligosaccharides, 75
Chitosan, 9, 18, 75, 101, 117, 118–128, 138, 139, 144, 145, 159
  A-deoxycholic acid nanoparticles, 123, 124
  chitosan composites, 127

collagen-alginate cushion, 70
gold nanoparticles, 126
microsphere/hydrogel scaffolds, 125, 126
molecule, 140
N-acetyl galactosaminyl transferase I & II, 119
nisin nanogel, 121
polymerization factor, 119
Chondrocytes, 107, 126, 145
Chondroitin, 117–120, 128, 144
  sulfate, 8, 75, 119, 121, 123, 128, 144
  glucuronyltransferase, 119
Ciprofloxacin-loaded gelatin microspheres, 71
Cisplatin, 15, 102
Coagulation, 43, 151, 156, 157
Coccinia grandis, 70
Collagen, 61–64, 66–86, 93–95, 104, 147–149, 158
  cell carrier (CCC), 148, 149
  extraction procedure from animals, 64
    collagen fibril, 66
    collagen, 64
    cross-linked collagen, 66
    tropocollagen, 64
  nanomaterial-drug hybrid scaffold, 69
  wound dressing materials, 69
Colorectal cancers, 83
Corneal collagen cross-linking (CXL), 72–74
Corrected distance visual acuity (CDVA), 74
Cortisol, 35
Cosmetic surgery, 62
Cosmetics, 19, 94, 97, 133, 136–138, 146
Covalent bonding, 36, 48
Crustaceans, 138
Cyamopsis tetragonolobus, 146
Cyclosporine, 8
Cytokine secretions, 148
Cytokines, 43, 148
Cytoplasm, 38, 47
Cytotoxic effects, 150
Cytotoxicity, 45, 48, 51, 70, 75, 96, 99, 126, 142, 144, 149, 153–155

D

Degradation, 1, 6, 15, 34, 38, 75–77, 83, 85, 97, 100, 125, 127, 139, 140, 147–153, 161
Dehydrothermal
  processing (DHT), 77
  treatment, 96
Deoxycholic acid, 123, 124
Dermis-like tissue area, 69
Desolvation technique, 99–101
Dextran, 146
D-galactan, 145
D-galactose units, 145
D-glucose molecules, 146
D-glucuronic
  acid, 1, 4, 119, 144
    transferases, 119
  residue, 117, 119
Diclofenac, 7, 8
Differential scanning calorimetry (DSC), 121, 127, 158
Diisopropyl fluorophosphate(DFP), 68
Disaccharide, 1, 4, 5, 144
Docetaxel, 48, 51
Dorsal root ganglia (DRG), 150
Doxorubicin, 43, 45, 48, 50, 100, 123, 124
  loaded albumin nanoparticles, 48
Dresden protocol (C-CXL), 72, 73
Drug
  affinity complex (DAC), 43, 51
  delivery, 2, 4, 6, 7, 10–12, 14, 16, 17, 19, 33, 34, 47, 50, 61, 62, 71, 93, 96–98, 100, 103, 106, 108, 109, 117, 123, 127, 128, 133, 135, 139, 141, 142, 144–147, 150, 158–161

E

Ehlers-Dalnos syndrome, 62
ELISA techniques, 139
Emulsification, 145
Emulsion
  evaporation technique, 98
  stabilization, 33, 34
Encapsulation, 48, 103, 124–126, 145, 148
Endocytosis, 38, 39, 124

Endogenous
  exogenous ligands, 35
  proteins, 34
Endoplasmic reticulum, 34
Endosomes, 40
Endothelial cells, 39, 72, 80, 84, 85, 105
Endothelium, 38
Endotoxin, 153
Enterococcus faecalis, 99
Enzymatic
  activities, 2, 159
  biodegradation, 136
  degradation, 148
  pre-treatment, 148
Enzyme
  controlled biodegradation, 136
  prolyl hydroxylase, 64
Enzymolysis, 15
Epidermal
  dermal layers, 71
  growth factor, 19, 51
    receptor 2 (EGFR2), 42
Epidermis, 8
Epidermolysis bullosa, 62
Epithelial cells, 10, 39, 48, 65, 71, 79
Erythropoietin, 16
Escherichia coli, 106, 157
Ethosomes, 9
Ethyl cellulose (EC), 138
Ethylene vinyl alcohol, 141
Eukaryotics, 157
European pharmacopeia (EP), 155
Extracellular matrix (ECM), 39, 64, 70, 78, 81–85, 104, 107, 108, 119, 134, 160, 161

F

Fatty acids, 35, 41
Femoral arteries, 79
Fermentation, 1, 4, 144
Ferromagnetic properties, 78
Fibrin, 151, 158, 159
Fibroblasts, 39, 69, 76, 105, 148
Fluorescence
  microscopy, 124
  polarization immunoassay, 12

Index 173

Food
   drug delivery, 33
   Fertilizer Technology Centre (FFTC), 139
Foot-and-mouth disease (FMD), 95
Fourier transform infrared (FTIR), 121, 122, 158
Furosemide, 37

G

Ganciclovir, 47
Gardenia jasminoides, 96
Gelatin, 3, 69–71, 93–109, 127, 149, 158
   graft copolymer, 98
   particles, 97
      liposomes, 103
      microparticles, 100
      nanoparticles, 97
Gene
   mutations, 149, 157
   therapy, 48, 159, 160
Genetic
   fusions, 33
   toxicology, 155
Genipa americana, 96
Genipin, 18, 80, 96, 149, 153, 154
Genotoxicity tests, 149, 157
Gentamicin, 12, 13
   microparticulates, 12
   sulfate, 107, 108
Glioblastoma, 100
Glucagon, 35
   like peptide-1 agonist (GLP-1), 41, 43, 44
Glucopyranosyl hydroxyl groups, 138
Glucosamine, 144
Glucosyl-galactosyl residues, 64
Glucuronic acid, 4, 37
Glutaraldehyde, 47, 96, 98, 101, 103, 149
Gly-A-B amino acid residue, 63
Glycoprotein receptors, 38
Glycoproteins, 38, 145
Glycosaminoglycan, 117–120, 128, 143, 144
Glycyrrhetinic acid receptor, 48

Gold nanoparticles, 126
Golgi bodies, 4
Gp60-albumin, 38
Granulation, 104, 139, 153, 159
Graphene oxide–polyethylene glycol (GO-PEG), 69
Guar kernels (*Cyamopsis tetragonolobus*), 146
Guinea pig maximization test (GPMT), 156
Gums, 145, 160

H

Heart failure (HF), 81, 82
Heavy chain (HC), 40
   antibody fragments, 41
Helical structure, 35, 62, 63, 65
Hematology, 156
Hemocompatibility, 70, 80, 134, 154, 155
Hemolytic effect, 157
Hemorrhage, 44
Hemostasis, 68
Heparin–collagen, 80
Hepatitis C, 43
Hepatocytes, 34, 157
Hepatoma, 48
Heterocyclic compounds, 35
Heterodimeric type I glycoprotein, 40
Heterogeneity, 149
Heterotrimers, 63
Hexamethylene-diisocyanate, 148
Hexosamine, 70
Histidine residues, 40
Histopathology, 156
Homopolymeric blocks, 142
Hormones, 33, 35, 43, 44, 160
Human
   gastric cancer cell line (HGC-27), 123
   intestinal epithelial cells, 40
   lymphocytes, 157
   serum albumin (HSA), 33–37, 42, 47, 48, 51
Hyaluronan, 4, 9, 11, 19, 136, 143, 144
   film, 9
   synthases (HAS), 4, 19

174   *Index*

Hyaluronic acid (HA), 1–19, 77, 78, 80, 106, 119, 142, 143, 153
  chitosan microparticulate system, 13
  diclofenac gel, 8
  gel, 7, 8
  hydrogel, 16
  hydroxyethyl cellulose hydrogels, 9
  microhydrogel, 16
  modified nanostructured lipid carriers, 9
  nasal delivery, 13
  paclitaxel
    conjugate, 15
    drug conjugate, 15
  poly(N-isopropyl acrylamide), 9
  polyethyleneimine-dexamethasone/DNA, 15
  polymer, 10, 14
  topical drug delivery, 8
Hyaluronidase, 1, 6, 15
Hydrogen bonds, 147, 152
Hydrolysis, 6, 15, 62, 93, 94, 138, 139, 152, 153
Hydrophobic
  drugs, 69, 97, 98, 104
  ester groups, 138
  interactions, 152
Hydrophobicity properties, 98
Hydrothermal methods, 78
Hydroxy prolyl residues, 64
Hydroxyethyl
  cellulose (HEC), 138
  methacrylate, 98
Hydroxyl
  group, 5, 15, 64
  lysyl amino acid, 64
Hydroxyproline, 63, 70, 93
Hydroxypropyl
  cellulose (HPC), 13, 138
  methylcellulose (HPMC), 138
Hypersensitivity, 2, 155
Hypertensive heart disease, 81
Hypertrophic scar formation, 148
Hypotonicity, 10
Hypovolemia, 146
Hypoxanthine-guanine phosphoribosyl transferase assay, 157

I

Ibuprofen, 8, 35, 37, 98, 106
Immune
  cells, 85, 148
  reactions, 139
  response, 84, 148, 150
  system, 38, 148
Immunogenicity, 34, 37
Immunohistochemical articulation, 77
Implantable
  collamer lens (ICL), 74
  drug delivery, 14
Insulin, 13, 14, 35, 41, 70, 126, 159, 160
Interstitial colloid osmotic pressure, 35
Intramolecular disulfide bonds, 152
Intramuscular (IM) injections, 12, 19, 155
Intravenous (IV) injections,, 12, 19, 43, 63, 65, 84, 85, 147, 156
Ionic bonds, 152
Irritation reactivity, 155
  intracutaneous reactivity, 156
Isoelectric point, 34, 36, 37, 97
Isoliquiritigenin, 9
Isotonic solution, 10

K

Kaolin, 106
Kappaphycus alvarezii, 146
Keloid fibroblast, 9
Keratinocyte, 69, 70, 76
Keratoconus, 72–74
Keratometric values, 73
Keratosis, 7
Kidney proximal tubules, 39

L

Laminaria hyperborean, 142
Leguminosae, 146
Lignocellulosic plants, 137
Liposomal formulations, 125
Liposomes, 9, 11, 17, 97, 103, 104, 124
Liver
  endothelium cell membranes, 38
  hepatocytes, 33
Lymphoblastic leukemia, 48

Index 175

Lymphocyte proliferation assays, 139
Lysine, 35, 41, 43, 63, 64, 105, 152
Lysozyme, 127, 139
Lysyl hydroxylase, 64

M

Macrocystis pyrifera, 142
Macromolecular
  component, 143
  functions, 6, 144
Macrophages, 38, 69
Mammalian
  cells, 157
    culture media (MEM), 155
  implantation model, 139
Marfan syndrome, 62
Matrix metalloproteinases (MMPs), 82,
  83, 148, 151
Megalin, 38, 39
  cubilin receptor, 39
Membrane
  forming materials, 138
  matrix, 107
Mesenchymal stem cells (MSC), 78, 105,
  143
Metalloproteinases, 83, 147, 148
Methotrexate, 43
Methylcellulose (MC), 138
Microbial flora absence, 146
Microcapsules, 100, 103, 143
Microcrystalline cellulose, 159
Microencapsulation, 93, 103
Microfluidic technique, 103
Micrometric
  matrix systems, 100
  reservoir systems, 100
Micronucleus assay, 157
Microparticulate system, 12, 13
Microscopic level, 83, 157
Molecular weight, 1, 5, 6, 11–13, 33, 34,
  37, 38, 43, 62, 126, 147
Monomeric
  gliadins, 152
  phosphoglycoprotein, 33, 37
  proteins, 152
Monosaccharides, 134, 144

Mouse lymphoma assay, 157
Mucoadhesive properties, 7, 8, 12
Mucoadhesiveness, 8
Myeloid leukemia, 48
Myocardial fibrosis, 81
Myristic acid, 41

N

N-acetyl
  D-galactosamine residue, 117, 119
  galactosamine, 119, 144
  glucosamine, 4
    groups, 138
N-ethylmaleimide (NEM), 68
N-terminal lysine, 41
N,N'-methylene bisacrylamide, 120, 121
Nanocarrier, 97, 98
Nanocomposite film, 98
Nanogabapentin, 47
Nanoparticle
  hyaluronate gel, 9
  system, 17, 98
Nanoparticles, 11, 17, 18, 45, 47, 48, 50,
  62, 70, 71, 93, 97–102, 123–127, 145,
  154, 159
Nasal
  administration, 142
  pulmonary drug delivery, 10
  route, 12, 13
  spray formulation, 13
Natural
  biochemical, 34
  biopolymers, 2, 3
  gelling agents, 94
  polymer, 2, 97, 109, 118, 133, 136, 146,
    147, 153, 158, 161
Necropsy, 156
Neoepithelial length, 69
Neointimal hyperplasia, 80
Neonatal
  FC receptor (FCRN), 39–41
  intestinal epithelial cells, 40
Neoplastic transformation, 149
Neutrophils, 139, 153
Non-collageneous (NC), 84

176   *Index*

Nonsteroidal anti-inflammatory drugs (NSAIDs), 37, 51
Nucleophilic attack, 36
Nucleo-shell structure, 15

O

Oligonucleotide, 99
Ophthalmic drug delivery, 8, 10, 11
Opthalmology, 71, 86
Oral administration, 126, 128, 139
Oropharyngeal aspiration, 51
Orthopedic
  imperfections, 76
  surgery, 6
Osmolarity, 10
Osmotic pump delivery processes, 159
Osteoarthritis, 117, 119, 128, 144
Osteoblast cells, 78
Osteoconductive framework, 76
Osteoconductivity, 77
Osteogenesis, 78
  imperfecta, 62
Osteogenic differentiation, 77, 107
Osteoid matrix, 77
Osteoinductive properties, 78
Osteoporotic fracture, 78
Ovalbumin (OVA), 33, 34, 37, 38, 127, 128
Ovarian cancer cells, 48, 50

P

Paclitaxel, 15, 45, 50, 104
  bound albumin nanoparticle, 45
Pancreatic cancer, 45, 99
Papain, 139, 149, 153
Parenteral injection, 43
Penicillins, 41
Pentylenetetrazole, 47
Pepsin soluble collagen, 67
pH-dependent albumin, 40
Pharmaceutical
  applications 2, 4, 7, 19, 34, 44, 47, 85, 104, 117, 128, 134–136, 146, 159, 160
    gelatin, 96
  excipient, 2, 128
  formulation, 135, 159

industry, 94, 108, 118, 133–135, 137, 145, 146, 158–160
  preparations, 135
Pharmacodynamics, 45, 51
Pharmacokinetic
  parameter, 45
  profile, 42
  properties, 42
Pharmacokinetics, 45, 51, 147
Pharmacotherapy, 97
Phenylmethanesulfonyl fluoride (PMSF), 68
Phosphorylation, 119
  reaction, 119
Photographic films, 138
Photorefractive keratectomy (PRK), 73
Photosensitizer solution, 72
Physicochemical properties, 3, 10, 77, 100, 157
Pichia pastoris, 44
Pigment dispersion syndrome (PDS), 73
Pilocarpine, 10, 11, 72
Plasma
  compartments, 34
  treatment, 96
Plasticizers, 157
Platelets, 80, 157
Poly-anionic biomacromolecular polysaccharide, 1, 3
Poly(3-hydroxybutyric acid)-gelatin, 70
Poly(lactide-co-glycolide) (PLGA), 75
Poly(ε-caprolactone) (PCL), 70, 107
Polyanion, 3
Polycaprolactone, 69, 142
Polydispersity index, 125
Polyepoxy compounds, 148, 149
Polyethylene glycol (PEG), 69, 98, 99, 105
Polyethyleneimine-dexamethasone/DNA, 15
Polylactic acid, 141
Polymeric
  chain, 118
  medical device, 153
Polypeptide
  chain, 35, 63, 147, 150–152
  drugs structure, 16
  sequences, 152

Index

177

Polysaccharide
  polymers, 146
  structure, 118
Polysaccharides, 3, 6, 64, 118, 125,
  134–136, 138, 140, 142, 144–146, 159,
  160
Polytetrafluoroethylene (PTFE), 79
Polyvinylpyrrolidone (PVP), 71
Porous scaffolds, 142
Post-translational modifications, 44, 64
Preformed conjugate-drug affinity
  complex (PC-DAC), 43
Procollagen type III N-terminal propeptide
  (PIIINP), 81, 82
Protease resistant
  derivative, 44
  GLP-1 receptor, 44
Protein, 4, 14, 16, 18, 33–41, 43, 44, 47,
  50, 51, 61, 62, 71, 74, 82, 84, 93, 94,
  98, 101, 108, 117–120, 127, 128, 134,
  139, 142, 144, 147, 149, 150, 152–154,
  158, 160
  biopolymers, 134
  delivery, 62, 128
  drugs, 16
  natural polymers, 118
  origin polymers, 147
  peptide drugs, 16
Proteoglycan structure, 118
Proteolysis, 83–85, 149, 152
  collagen, 84
Proteolytic action, 84
Pseudomonas aeruginosa, 99, 108
Pulmonary
  delivery, 13, 18, 93
  diseases, 10
Pyrogenicity, 155, 156

Q

Quercetin, 69

R

Radical scavenging effect, 44
Radioisotope, 10
Rapamycin, 48

Red blood cells, 157
Redox regulators, 43
Regeneration, 3, 62, 64, 71, 75, 76, 85,
  106, 107, 134, 140, 142, 145
Response surface methodology (RSM),
  121
Retrogradation tendency, 141
Riboflavin, 72, 73
Ribosomes, 64
Rifampicin, 98, 103, 104

S

Saccharomyces cerevisiae, 44
Salicylic acid, 37, 47
Salmonella typhimurium, 157
Scanning electron microscopy (SEM), 121
Scavenging activity, 6
Schwann cells, 150
Secreted protein, acidic, and rich in
  cysteine (SPARC), 38
Silk fibroin, 75, 150, 151
Single-chain antibody fragment (scFv), 41
Skin
  aging, 62, 75
  infection, 7
  model, 74
  replacement, 74, 75
  sensitization, 103
  synovial fluid, 4
  treatment, 158
Smooth muscle cells (SMCs), 80
Sodium
  carboxymethyl cellulose (NaCMC), 8,
    138
  hyaluronate, 10–13, 18
    microparticles, 13
    solutions, 12
  tetrathionate, 16
Solvent
  evaporation technique, 12
  taxane formulations, 45
Soy protein, 152
  biodegradable films, 153
Soya-lecithin based liposomes, 104
Sponges, 62, 69, 86, 145

178 *Index*

Squamous cell carcinoma, 15
Staphylococcus
   aureus, 106
   epidermidis, 108
Starch, 140, 160
   cellulose acetate (SCA), 142
   ethylene vinyl alcohol (SEVA-C), 142
   materials, 141
   polymers, 141, 142
   polyvinyl alcohol, 153
Sterculia
   gum, 3, 146
   urens, 146
Stickler syndrome, 62
Streptozotocin rat, 126
Succinimidyl ester polyethylene glycol, 148
Sulfation mechanism, 119
Sulfhydryl groups, 38
Synovial fluid, 1, 3–6, 118, 144

T

Taxanes, 15
Therapeutic
   agent, 7
   efficacy, 40
   peptides, 40
Thermogravimetry analysis (TGA), 121, 127
Thrombosis, 79, 80, 157
Thyroid, 35
Tissue
   engineered matrices, 107
   engineering, 61, 71, 76, 93, 96, 105, 108, 117, 126, 128, 139, 141–143, 145, 147, 150, 158, 159
   inhibitors of matrix metalloproteinases (TIMP 9), 83
Titanium dioxide, 71
Topical
   drug delivery, 7
   formulation, 7
Topography-guided corneal collagen cross-linking (TG-CXL), 73

Toxicity, 2, 15, 16, 37, 45, 47, 48, 75, 76, 123, 136, 138, 139, 142, 146, 147, 149, 156
Tracheomalacia, 79
Tranexamic acid, 106
Transcytosis, 38
Transdermal drug delivery, 8, 9
Transmission electron microscopy (TEM), 122
Trichloroacetic acid, 68
Trypsin, 149, 153
Tryptophan, 35, 37, 51
Tumor
   associated collagen signatures (TACS), 83
   cells, 15, 47, 83–86
   necrosis factor-α (TNF-α), 42
   progression, 83–85
   tissues, 37, 43, 48
Type-B gelatin, 99

U

Ultraviolet
   treatment, 96
   UA light, 73
   UVA light, 72
Umbilical cord, 1, 3–5
Uncorrected distance visual acuity (UDVA), 74
United States Pharmacopeia (USP), 155
Urinary
   incontinence, 80
   obstruction, 80
Uronic acid, 1, 4, 70

V

Vaccines, 98, 159, 160
Vasopressin, 12
Viremia effect, 104
Viscoelastic, 2, 3, 7
Viscosity, 6, 8, 18, 42, 145
Viscous solutions, 12
Vitreous humor, 1, 3, 4, 6, 65, 144

Index

179

W

Warfarin, 35, 41
  azapropazone binding site, 51
Waste
  purification, 138
  solubility, 145
  water treatment, 136, 139
Wheat gluten, 152
Wound, 6, 9, 62, 63, 68–71, 85, 104–108,
  134, 139, 150, 159
  healing, 9, 62, 68–71, 85, 104–108, 134,
    139, 142
  model, 70

X

Xenograft models, 45, 47, 48
X-ray
  crystallographic examinations, 35
  crystallography, 40
  diffraction (XRD), 12, 127

Z

Zidovudine, 98
Zinc
  ions, 13
  oxide, 71, 72, 99